PROTECTING HOMELAND SECURITY; A STATUS REPORT ON INTEROPERABILITY BETWEEN PUBLIC SAFETY COMMUNICATIONS SYSTEMS

HEARING

BEFORE THE

SUBCOMMITTEE ON TELECOMMUNICATIONS AND THE INTERNET

OF THE

COMMITTEE ON ENERGY AND COMMERCE

HOUSE OF REPRESENTATIVES

ONE HUNDRED EIGHTH CONGRESS

SECOND SESSION

JUNE 23, 2004

Serial No. 108–98

Printed for the use of the Committee on Energy and Commerce

Available via the World Wide Web: http://www.access.gpo.gov/congress/house

U.S. GOVERNMENT PRINTING OFFICE

95–444PDF   WASHINGTON : 2004

COMMITTEE ON ENERGY AND COMMERCE

JOE BARTON, Texas, *Chairman*

W.J. "BILLY" TAUZIN, Louisiana
RALPH M. HALL, Texas
MICHAEL BILIRAKIS, Florida
FRED UPTON, Michigan
CLIFF STEARNS, Florida
PAUL E. GILLMOR, Ohio
JAMES C. GREENWOOD, Pennsylvania
CHRISTOPHER COX, California
NATHAN DEAL, Georgia
RICHARD BURR, North Carolina
ED WHITFIELD, Kentucky
CHARLIE NORWOOD, Georgia
BARBARA CUBIN, Wyoming
JOHN SHIMKUS, Illinois
HEATHER WILSON, New Mexico
JOHN B. SHADEGG, Arizona
CHARLES W. "CHIP" PICKERING,
  Mississippi, *Vice Chairman*
VITO FOSSELLA, New York
STEVE BUYER, Indiana
GEORGE RADANOVICH, California
CHARLES F. BASS, New Hampshire
JOSEPH R. PITTS, Pennsylvania
MARY BONO, California
GREG WALDEN, Oregon
LEE TERRY, Nebraska
MIKE FERGUSON, New Jersey
MIKE ROGERS, Michigan
DARRELL E. ISSA, California
C.L. "BUTCH" OTTER, Idaho
JOHN SULLIVAN, Oklahoma

JOHN D. DINGELL, Michigan
  Ranking Member
HENRY A. WAXMAN, California
EDWARD J. MARKEY, Massachusetts
RICK BOUCHER, Virginia
EDOLPHUS TOWNS, New York
FRANK PALLONE, Jr., New Jersey
SHERROD BROWN, Ohio
BART GORDON, Tennessee
PETER DEUTSCH, Florida
BOBBY L. RUSH, Illinois
ANNA G. ESHOO, California
BART STUPAK, Michigan
ELIOT L. ENGEL, New York
ALBERT R. WYNN, Maryland
GENE GREEN, Texas
KAREN McCARTHY, Missouri
TED STRICKLAND, Ohio
DIANA DeGETTE, Colorado
LOIS CAPPS, California
MICHAEL F. DOYLE, Pennsylvania
CHRISTOPHER JOHN, Louisiana
TOM ALLEN, Maine
JIM DAVIS, Florida
JANICE D. SCHAKOWSKY, Illinois
HILDA L. SOLIS, California
CHARLES A. GONZALEZ, Texas

BUD ALBRIGHT, *Staff Director*
JAMES D. BARNETTE, *General Counsel*
REID P.F. STUNTZ, *Minority Staff Director and Chief Counsel*

SUBCOMMITTEE ON TELECOMMUNICATIONS AND THE INTERNET

FRED UPTON, Michigan, *Chairman*

MICHAEL BILIRAKIS, Florida
CLIFF STEARNS, Florida
  Vice Chairman
PAUL E. GILLMOR, Ohio
CHRISTOPHER COX, California
NATHAN DEAL, Georgia
ED WHITFIELD, Kentucky
BARBARA CUBIN, Wyoming
JOHN SHIMKUS, Illinois
HEATHER WILSON, New Mexico
CHARLES W. "CHIP" PICKERING,
  Mississippi
VITO FOSSELLA, New York
STEVE BUYER, Indiana
CHARLES F. BASS, New Hampshire
MARY BONO, California
GREG WALDEN, Oregon
LEE TERRY, Nebraska
JOE BARTON, Texas,
  (Ex Officio)

EDWARD J. MARKEY, Massachusetts
  Ranking Member
ALBERT R. WYNN, Maryland
KAREN McCARTHY, Missouri
MICHAEL F. DOYLE, Pennsylvania
JIM DAVIS, Florida
CHARLES A. GONZALEZ, Texas
RICK BOUCHER, Virginia
EDOLPHUS TOWNS, New York
BART GORDON, Tennessee
PETER DEUTSCH, Florida
BOBBY L. RUSH, Illinois
ANNA G. ESHOO, California
BART STUPAK, Michigan
ELIOT L. ENGEL, New York
JOHN D. DINGELL, Michigan,
  (Ex Officio)

CONTENTS

Page

(III)

PROTECTING HOMELAND SECURITY; A STATUS REPORT ON INTEROPERABILITY BETWEEN PUBLIC SAFETY COMMUNICATIONS SYSTEMS

WEDNESDAY, JUNE 23, 2004

House of Representatives,
Committee on Energy and Commerce,
Subcommittee on Telecommunications
and the Internet,
Washington, DC.

The subcommittee met, pursuant to notice, at 1:39 p.m., in room 2123, Rayburn House Office Building, Hon. Fred Upton (chairman) presiding.

Members present: Representatives Upton, Gillmor, Cox, Shimkus, Fossella, Bass, Terry, Barton (ex officio), Wynn, McCarthy, Stupak, and Engel.

Staff present: Will Nordwind, majority counsel and policy coordinator; Howard Waltzman, majority counsel; William Carty, legislative clerk; and Peter Filon, minority counsel.

Mr. UPTON. Must be 1:30. Good afternoon. You might say that I know that there are a number of subcommittees this afternoon, and I also know that late yesterday we were notified that Secretary Powell is going to be up briefing members on the situation in Iraq in a few minutes, so I would think that a number of members will be torn when it comes to where they are going to appear. We are in session as well with votes expected in the not too distant future as well.

But good afternoon. Today's hearing is entitled, "Protecting Homeland Security: A Status Report on Interoperability Between the Public Safety Communications System," and it represents this subcommittee's continuing effort to examine matters related to homeland security within its jurisdiction.

Sadly, we live in dangerous times. Since 9/11, our Nation has tried to learn from the bitter events of the past and better prepare to respond during the next crisis, whether it is, God forbid, another terrorist attack, natural disaster, a criminal act or something like the blackouts of last summer.

In all such events, it is our Nation's first responders who answer the call of duty. As citizens flee and evacuate to protect themselves, our Nation's first responders are running the opposite direction, usually into harm's way. Nowhere was this selfless service more self-evident than in Ground Zero on 9/11. To paraphrase Admiral Nimitz on that day, "Our Nation's first responders showed us that

(1)

uncommon valor was a common virtue." But I would submit that what we citizens consider uncommon valor, our Nation's first responders humbly consider to be just doing their jobs.

But in order to better equip them to do their jobs, our Nation's first responders need to be able to communicate on their radios with one another, not only in terms of fire, police, EMS within a jurisdiction but also amongst local, State and Federal jurisdiction. However, achieving interoperability throughout our Nation has proved to a monumental and multifaceted challenge.

Today, we will hear about some of those challenges are being confronted and the status of progress being made throughout our Nation. I am particularly pleased with the leadership demonstrated by the FCC, Department of Homeland Security, as they work with stakeholders at all levels of the government and communities around the country to achieve interoperability. As we will hear today from some of our witnesses, one critically important challenge is to ensure that public safety has the spectrum that it needs in bands which are well suited for interoperability.

Back in 1997, Congress directed 24 megahertz for spectrum in the upper 700 megahertz to be allocated to public safety. However, the spectrum is encumbered by broadcasters and will be until the transition to digital TV is complete. That is why this subcommittee has spent an enormous amount of time working to ensure the expeditious completion of the digital TV transition.

Several weeks ago, this subcommittee examined a proposal by the FCC's Media Bureau which would, in effect, provide a clear path to the completion of the transition. There are many sound policy reasons to pursue that plan, not the least of which is getting public safety the spectrum that it needs to achieve interoperability. As such, this subcommittee will vigorously continue its work to advance the digital TV transition in the months to come.

So today I look forward to hearing from our witnesses about how our Nation is progressing toward interoperability and without a doubt interoperability is a critical necessity for our Nation's first responders as they help protect our homeland security.

I would yield to my colleague, the gentleman from California, Mr. Cox, for an opening statement.

Mr. COX. Thank you very much, Mr. Chairman. These are very important hearings. As you point out, the terrorist attacks on September 11, 2001 and everything that we have been facing since demonstrates on a continuing basis the vital need for interoperable communications among our first responders.

On the Homeland Security Committee, which I chair—is this mike a problem? This opening statement is just electric.

Is that the sound of one or two hands clapping? All right. On the Homeland Security Committee, which I chair, we have been working to get this problem of interoperable communications solved, but the truth is the jurisdiction to do that in the most effective way lies with this committee, the Energy and Commerce Committee, and that makes the testimony that our witnesses are going to present today of special importance.

From the first day in March 2003 that it was in operation, the Department of Homeland Security took the lead in managing Project SAFECOM, a Federal program established by President

Bush in 2001 to help local, State and Federal agencies improve emergency response through interoperable wireless communications.

The good news is that DHS will contribute $21 million to this joint effort in fiscal 2005. That is a near doubling of the current funding level. And based on the President's proposed budget, other agencies, including the Departments of Defense, Energy and HHS, as well as the Department of Justice, will contribute an additional $10 million to this joint effort.

Since the creation of SAFECOM, and particularly in the last year, under DHS leadership, we have seen real progress in this area, as Dr. Boyd will describe in detail. The department has developed guidance for public safety interoperability equipment grants, and in April 2004, it established the first ever set of interoperability requirements. These will help our Nation's first responder community in determining their interoperability needs.

The Secretary also is establishing a separate office within the S&T directorate to manage and oversee issues relating to interoperability and compatibility, including Project SAFECOM. Most important, the department recently announced a short-term incident level interoperable communications strategy to tackle interoperability problems in 10 of America's highest risk cities.

At relatively low cost, first responders will be able to interact by voice with each other regardless of frequency or mode. They will use a patching or a bridging network set up to manage the specific incident. This will ensure that we have an extra layer of interoperability protection now while we continue to work toward resolving the technical and cost issues relating to the more long-term solutions.

Since 9/11, this Congress and this administration have made billions of dollars available to State and local governments to purchase interoperable communications through our terrorism preparedness grant programs at DHS and at the Department of Justice. The technical issues and the lack of standards have prevented quick acquisition of such technology by the first responder community.

I am pleased that the Department of Homeland Security is moving aggressively to address both the short-term needs of our high-risk areas and, in conjunction with other Federal agencies, the long-term challenges of technology development, standard setting and spectrum limitations.

And that takes us to the central issue of spectrum. I would like to commend Chairman Barton for his outstanding recent comments affirming the need to reclaim the analog TV spectrum by the end of 2006. These multibillion dollars slices of the airwaves are now being used to send duplicate TV signals over the air, two identical signals from each station, soaking up the most valuable of our airwaves, even as the population of TV viewers who actually receive their programming over the air continues to decline.

Mr. Grube from Motorola will perform the valuable public service today of describing for us exactly how small this audience is. Perhaps we can go further and inquire how many of the folks in this small audience simply cannot afford either cable or satellite service or a $100 digital analog converter and how many of these folks just don't care about television.

Maybe if we gave them the choice, they would choose super high speed wireless Internet access or less expensive cell phone service or a broadband public safety network or the next generation of telemedicine. Maybe they would choose any or all of these things over the right to watch endless reruns of "Saved by the Bell" without upgrading the receiver.

When we talk about the huge swaths of our public airwaves that have been given away for nothing to the broadcasting industry, we usually think of the billions lost to the taxpayers who are forced to underwrite this subsidy or the lost consumer opportunities when new wireless applications are starving for bandwidth. But as we will hear today, lack of available spectrum can also impose human costs that are beyond measure.

I want to thank our witnesses for their testimony today. I want to thank Chairman Upton and everyone on the staff and in the audience who is working on this problem for your tireless efforts to bring interoperable communications technology to our first responders.

Again, Thank you, Mr. Chairman.

Mr. UPTON. Thank you.

Mr. Shimkus?

Mr. SHIMKUS. Thank you, Mr. Chairman. I would just focus on welcoming a good Illinois company, Motorola, to the panel and say that when members—for members to be successful, sometimes we have to specialize just like anything else. And I have been fortunate to be involved with the E911 systems, 911 as part of an area that I have tried to focus on.

We have legislation that passed this committee and went to the floor, and now we await Senate action on the House version of the bill, which is H.R. 2898, trying to move and get all the stakeholders engaged in having a true enhanced 911 system that is also positioned—a location identification through cellular systems and by GPS chips.

The importance in homeland security and other issues is what we also find out in hearings, is the ability for the public agencies to, in essence, to recall from the signals or call to the signals to warn people in areas in which there may be a biological attack and the wind drift areas and the like.

So I will use this to continue to promote movement on that bill and encourage individuals to work with our Senate colleagues to make sure that Enhanced 911 is going to get enacted, and we get legislation passed on the Senate side and we get reconciliation and we get a bill that the President can sign, it is very important in this whole debate.

And with that, Mr. Chairman, thank you, and I yield back my time.

[Additional statements submitted for the record follow:]

PREPARED STATEMENT OF HON. PAUL E. GILLMOR, A REPRESENTATIVE IN CONGRESS FROM THE STATE OF OHIO

I thank the Chairman for the opportunity to examine the ability of public-safety agencies to seamlessly communicate with each other. In an environment of terrorist threats, criminal activity, natural disasters, and everyday-life emergencies, it is imperative that we address such issues and continue lay the groundwork to ensure that first responders have the tools they need to keep us safe, healthy, and secure.

Interoperability is certainly achievable, and I applaud the parties represented today who have made great strides in overcoming different coordination, technology, budget, and frequency challenges faced by local, state, and federal agencies across the country.

I look forward to hearing from each of the witnesses, and in particular the progress being made from such initiatives as Safecom, but also other examples of current difficulties in achieving interoperability as well as a prospective timeline as to when our local, state, and federal agencies will be able to respond in a more coordinated and consistent manner during an urgent situation, wherever and whatever it may be.

Again, I thank the Chairman for bringing attention to this important issue and yield back the remainder of my time.

PREPARED STATEMENT OF HON. JOE BARTON, CHAIRMAN, COMMITTEE ON ENERGY AND COMMERCE

Mr. Chairman, thank you for calling this hearing today. Protecting our homeland security is a top priority of this committee, and I applaud you for holding a hearing to examine the progress being made in ensuring that our nation's emergency communications systems are interoperable.

It is critical that first responders be able to communicate before, during, and after terrorist attacks, criminal acts, and natural disasters. The notion that police and fire departments from the same city cannot use their handheld radios to communicate with each other is mindboggling. And it is no less surprising that local officials cannot use their radio systems to communicate with state or federal officials.

Part of the problem stems from the fact that first responders use disparate frequencies for their communications systems. That is why it is so important for television broadcasters to return spectrum in the Upper 700 MHz band currently used to provide analog television service. In 1997, Congress identified 24 MHz of spectrum in this band for public safety use. However, until the broadcasters vacate the band, the spectrum is virtually worthless to public safety. As a result, Congress needs to enact a hard date for the digital television transition so that the broadcasters vacate the band.

The 24 MHz allocated in the 700 MHz band is ideal for interoperability. First responders across the nation could use this spectrum to share common channels on which multiple local, state, and federal agencies could coordinate emergency response.

Achieving interoperability between emergency communications systems will save lives. I appreciate the efforts being undertaken by the Department of Homeland Security and the FCC to make interoperability a reality. I encourage these agencies to continue to work within the Executive Branch as well as with state and local officials and industry to make every community's communications systems ready to prevent or mitigate a possible terrorist attack.

Mr. Chairman, thank you again for holding this hearing. I look forward to working with you and our colleagues to ensure that this committee is doing everything possible to protect our homeland security.

Mr. UPTON. Thank you. Well, today we are fortunate to have the witnesses that we have. And we will start with Dr. David Boyd, Deputy Director of the Office of Systems Engineering and Development at the Department of Homeland Security; Mr. Gary Grube, corporate VP and CTO, Commercial, Government and Industrial Solutions for Motorola; Mr. Robert LeGrande, deputy chief technology officer of the Office of the Chief Technology Officer of the District Of Columbia; and Mr. John Muleta, Bureau Chief of the Wireless Telecommunications, obviously from the FCC.

And, gentlemen, we appreciate your testimony. It will be made as part of the record in its entirety and we would like you to spend 5 minutes now, starting with Dr. Boyd, at which point when you are finished we will have questions from members of the panel.

Dr. Boyd, welcome.

STATEMENTS OF DAVID G. BOYD, DEPUTY DIRECTOR, OFFICE OF SYSTEMS ENGINEERING AND DEVELOPMENT, DEPARTMENT OF HOMELAND SECURITY; GARY GRUBE, CORPORATE VICE PRESIDENT AND CTO, COMMERCIAL, GOVERNMENT AND INDUSTRIAL SOLUTIONS, MOTOROLA INC.; ROBERT LEGRANDE, DEPUTY CHIEF TECHNOLOGY OFFICER, OFFICE OF THE CHIEF TECHNOLOGY OFFICER, DISTRICT OF COUMBIA; AND JOHN B. MULETA, BUREAU CHIEF, WIRELESS TELECOMMUNICATIONS, FEDERAL COMMUNICATIONS COMMISSION

Mr. BOYD. Well, thank you, Mr. Chairman, and good afternoon, members. Thank you for the invitation to speak to you today.

Earlier this year, Secretary Ridge observed that, "The ability of our first responders to communicate with each other as well as share equipment in times of crisis is a critical issue facing our Nation. Solving this challenge is a long-standing and complex problem. There are, however, some immediate steps the department can take this year to address the communications and equipment needs of first responders and make substantial progress to achieving the penultimate communications solution." To address the needs identified by emergency response providers, the Secretary has directed the establishment of intradepartmental program offices to address several key homeland security priorities. One of these is a program office to significantly improve the coordination and validation of the department's interoperability programs, thus allowing firefighters, police officers and other emergency personnel to better communicate and share equipment with each other during a major disaster.

The directorate of Science and Technology within DHS has been tasked to lead the planning and implementation of this office in coordination with other DHS programs. We recognize that for this office to succeed, emergency response providers and homeland security practitioners who own, operate and maintain more than 90 percent of the Nation's wireless public safety infrastructure must be integrated into the program from its beginning, so the solutions we create are solutions that will actually meet their needs. Cooperation and coordination with existing programs is key to reducing the necessary duplication of effort and allowing the leveraging of investment many public safety agencies have already made.

Properly designed, non-proprietary open architecture standards will maximize competition across industry, encourage technology innovation, reduce costs and help to ensure compatibility among public safety and homeland security agencies. Compliance with the National Incidence Management System, the National Response Plan and relevant homeland security Presidential directives will provide a consistent nationwide approach for agencies at all levels of government to work together to prepare for, prevent, respond to and recover from major incidents.

And, finally, outreach efforts will emphasize the need for interoperability and provide access to tools for its implementation. Initial priority portfolio areas that the office will address include communications to the most mature of the portfolios, equipment training and others as required.

We will model this office after the successful SAFECOM Program, which as a public safety practitioner-driven program works with existing Federal initiatives and key public safety stakeholders to address the development of better technologies and processes for the cross-jurisdictional and cross-disciplinary coordination of existing systems and future networks. We will do the same across all the portfolios for the more than 50,000 local and State public safety agencies and organizations and over 100 Federal agencies engaged in public safety disciplines, such as law enforcement, fire fighting, public health and disaster recovery.

The SAFECOM Program, which will continue as a key national initiative within the new Interoperability Office, has already made significant progress at achieving both its short-term goals and in building the foundations for a long term, comprehensive program. In fiscal year 2003, SAFECOM developed common grant guidance which was incorporated into the grant programs of the COPS Office, FEMA and ODP and which constituted the first coordinated effort to coordinate and align funding for communications programs in the Nation.

We published a comprehensive statement requirements for wireless public safety communications and interoperability in coordination with the National Public Safety Telecommunications Council, the major public safety associations, NIST, and the Department of Justice. The requirements identified in this document will drive the development and creation of interface standards needed to satisfy the needs of State and local responders. It offers industry the information they need to align their product development efforts with actual users' needs, and it will guide research, development, test and evaluation programs.

It constitutes the first national definition of what interoperability must accomplish, and within a month of its publication more than 5,000 copies were downloaded by public safety agencies, practitioners and manufacturers, and many of those manufacturers have already approached us to show us how they are mapping their capabilities to those requirements.

We will employ a system engineering or life cycle approach to identifying, defining and developing action plans in each portfolio area. Common components of this life cycle approach include the validation and means assessments, the development with the user community of a comprehensive statement of requirements for each portfolio, completion of baselines to provide starting points for each portfolio, a robust research and development program, a robust standards program to identify and adopt existing effective standards and to support the development of essential standards when none exist, testing and evaluation of technologies, development of appropriate grants and funding guidance and development of policy and legal reference materials or recommendations relevant to each portfolio.

To ensure that the efforts of this office are well coordinated an Interagency Interoperability Policy Board will be established to help reduce duplication in programs and activities. By the direction of Secretary Ridge, this new office has already undertaken a major initiative to achieve near-term emergency incident level interoperability in high-threat urban areas before the end of this year.

Working with a wide range of Federal agencies, including the DHS Office for Domestic Preparedness, the Justice Department and the National Guard, we have begun working with all 10 urban areas to identify what is already in place, what is available and what is still needed to provide interoperability to support a major incident.

As a Nation, we must continue to pursue a comprehensive strategy that takes into account technical and cultural issues associated with improving communications and interoperability. It must address research, development, testing and evaluation; procurement planning; spectrum management, including solving the current 800 megahertz interference problems and identifying and freeing up additional spectrum; standards training and technical assistance. And it must recognize the challenges associated with incorporating legacy equipment and practices in the face of a rapidly changing technology environment.

The many obstacles facing public safety interoperability makes for a complex interlocking set of problems with no one-size-fits-all solution. The new office, in company with a broad array of partners from all levels of government, is working toward a world where lives and property are not lost because public safety agencies are unable to communicate or lack compatible equipment and training resources.

I would be happy to answer any questions.

[The prepared statement of David G. Boyd follows:]

PREPARED STATEMENT OF DAVID G. BOYD, DIRECTOR, SAFECOM PROGRAM OFFICE, DIRECTORATE OF SCIENCE AND TECHNOLOGY, DEPARTMENT OF HOMELAND SECURITY

Good morning and thank you, Mr. Chairman and Members of the Committee for the invitation to speak to you today. I appreciate your interest in the Department's interoperability efforts and am grateful for this opportunity to address the important issue of public safety interoperability and compatibility before you.

PUBLIC SAFETY BACKGROUND

As Secretary Ridge stated on February 24, 2004,

> The ability for our nation's first responders to communicate with each other as well as share equipment in times of crisis is a critical issue facing our nation. Solving this challenge is a long-standing and complex problem. There are, however, some immediate steps the department can take this year to address the...communications and equipment needs of first responders and make substantial progress to achieving the penultimate communications solution.

Communications interoperability is the ability of public safety agencies to talk across disciplines and jurisdictions via radio communications systems, exchanging voice and/or data with one another on demand, in real time, as authorized. The nation is heavily invested in an existing infrastructure that is largely incompatible. Currently, efforts within the Federal government to address the interoperability problem are being coordinated to incorporate the needs of local, state, and Federal practitioners. However, there remain many challenges, both technical and cultural, facing the improvement of public safety communications and interoperability.

Whether fighting a fire or responding to a terrorist attack, efficient and effective emergency response requires coordination, communication, and the sharing of vital information and equipment among numerous public safety and security agencies. As the *National Strategy for the Physical Protection of Critical Infrastructures and Key Assets* makes clear, "systems supporting emergency response personnel, however, have been specifically developed and implemented with respect to the unique needs of each agency. Such specification complicates interoperability, thereby hindering the ability of various first responder organizations to communicate and coordinate resources during crisis situations."[1] The Department of Homeland Security (DHS or the Department) believes this issue is so important that it has identified interoperability of communications and equipment as the number two priority for the second year strategic plan. We seek to ensure the interoperability of critical emergency response systems or products by making it possible for them to work with other systems or products without special effort on the part of the user.

The Department also has developed intradepartmental program offices to address the needs identified by emergency response providers[2] and to respond to the problems identified in the *National Strategy for the Physical Protection of Critical Infrastructures and Key Assets*. One of these is a program office to significantly improve the coordination and validation of the Department's interoperability programs, thus allowing firefighters, police officers and other emergency personnel to better communicate and share equipment with each other during a major disaster.

Since its beginning, the Department has been involved with the issue of wireless interoperability through project SAFECOM. As a *public safety practitioner driven program*, SAFECOM, housed within the Department, has been the Federal government's central point to coordinate Federal wireless investments and activities and partner with State, local, and Tribal governments to improve the interoperability of our nation's wireless communications.

Secretary Ridge has now specifically tasked the Directorate of Science and Technology (S&T) within DHS, in coordination with other DHS programs, to lead the planning and implementation of an office of interoperability that will address the larger issue of interoperability, including wireless communications. By coordinating and leveraging the vast range of interoperability programs and related efforts spread across the Federal government, this office, currently titled the "Office of Interoperability and Compatibility" (OIC), will reduce unnecessary duplication in programs and spending and ensure consistency across federal activities related to research and development, testing and evaluation (RDT&E), standards, technical assistance, training, and grant funding related to interoperability. This new program office will encompass the SAFECOM office, which will continue as a key national initiative, into the effort to address the larger issue of interoperability.

Portfolio Areas

Within the OIC, we will create a series of portfolios to address critical interoperability and compatibility issues related to the emergency response provider and homeland security communities. Initial priority portfolio areas that the OIC will address, in coordination with other Departmental offices, including the DHS Office for Domestic Preparedness (ODP), include:

- Communications (through the SAFECOM Program Office);
- Equipment;
- Training; and
- Others as required.

To establish these portfolios, the OIC currently is identifying the necessary stakeholders and will utilize these stakeholders to assess and finalize the portfolio areas. Through this process, the OIC will identify the current initiatives as well as the most appropriate short-term deliverables.

Office Implementation

The OIC is being modeled after the SAFECOM Program, which represents a successful model for how to address highly sophisticated technical and policy issues associated with public safety communications and interoperability. Leveraging the work that the SAFECOM Program has already undertaken, the OIC will look to replicate relevant elements of the SAFECOM process and to build on SAFECOM's achievements in bolstering public safety communications.

The new OIC will employ a systems engineering or lifecycle approach to identifying, defining, and developing action plans in each portfolio area. This lifecycle approach is both iterative and collaborative. It emphasizes the need to create a common set of standards, policies, and procedures that encourage backwards compatibility of new solutions which will drive the migration of systems towards advanced, interoperable equipment and processes in the future. Common components of this lifecycle approach include:

---

[1] "National Strategy for the Physical Protection of Critical Infrastructures and Key Assets," The White House, February 2003, page 43.

[2] As defined in the Homeland Security Act of 2002, Section 2(6), "The term 'emergency response providers' includes Federal, State, and local emergency public safety, law enforcement, emergency response, emergency medical (including hospital emergency facilities), and related personnel, agencies, and authorities." 6 U.S.C. 101(6)

- Validation of needs assessments (consistent with Homeland Security Presidential Directive-8, which lays out the National Preparedness Goal, as appropriate);
- Development, with the user community, of a comprehensive statement of requirements for each portfolio;
- Completion of baselines to provide starting points for each portfolio;
- A robust research and development program for new capabilities;
- A robust standards program to identify and adopt existing, effective standards and to support the development of essential new standards when none exist;
- Testing and evaluation of existing technologies;
- Development of common standards for training and technical assistance;
- Development of appropriate grants/funding guidance; and
- Development of policy and legal reference materials or recommendations relevant to each portfolio.

Within the OIC, we are following the successful SAFECOM model by creating action plans for each of these areas, and for others identified as the portfolios are developed. Each of these action plans will be developed through a collaborative process that brings together the relevant stakeholders to provide clear direction on a path forward. The process to develop action plans will involve:

- Assessment of the government agencies involved in each portfolio;
- Identification of the relevant stakeholders at the local, state, and federal levels;
- A stakeholder working session to define the issues, assess user needs, and create a detailed vision of the "end state" for each portfolio; and
- A governance structure that ensures ongoing participation on the part of key stakeholders at the local, State, and Federal levels.

Through this end-user input, the new OIC will produce a strategy and action plan to address the interoperability and compatibility needs in each of these portfolios.

The OIC structure should be an organizational reflection of the lifecycle process it is designed to manage and support. The main purpose of the OIC will be to provide common standards of practice, protocol, planning, and evaluation across the broadest spectrum of interoperability activities and to facilitate the prioritization and coordination of these efforts within the framework of a common, nationwide vision.

The OIC will include a program management office responsible for coordinating the various portfolio managers and their respective management offices. In addition, a cross-departmental coordinating council or interagency interoperability policy board, chaired by the Undersecretary for S&T, will be established to ensure that its efforts are coordinated intra- and inter-departmentally. This board will help reduce duplication in programs and activities.

With respect to specific task, the new OIC has already, at the direction of the Secretary of Homeland Security, undertaken a major initiative—RapidComm 9/30—to achieve near term, emergency, incident-level interoperability in ten high threat urban areas by September 30, 2004. Working with a wide range of Federal agencies, including the ODP, the Justice Department, and the National Guard, we have begun working with all ten urban areas to identify what it is in place, what is available, and what is still needed to provide interoperability to support a major incident.

Players: Owners, Partners, and Stakeholders

Those with a vested interest in the OIC are the people, agencies, and organizations that will directly benefit from enhanced interoperability of equipment and processes. Creating interoperability requires coordination and partnerships among office managers, partners, and stakeholders. Secretary Ridge has directed that S&T will be the manager—or owner—of this office, and it will be essential for the office to establish partnerships with all relevant offices and agencies in order to effectively coordinate like-topic activities. These partners will be instrumental in ensuring that our programs address all possible issues, ranging from grants for equipment procurement to regulatory policy creation. These partners and additional relevant stakeholders include representatives from the following communities:

- Emergency response providers represented by their national associations;
- Department of Homeland Security and other government agencies
    - a. Operational programs and offices
    - b. Research & development offices
    - c. Test & evaluation programs
    - d. Technical assistance providers
    - e. Grant programs;
- Standards Development Organizations; and
- Industry

Principles for Achieving Interoperability

In order for the OIC to effectively coordinate and validate the Department's interoperability programs, it will need to employ a common set of standards, policies, and procedures. This will require that the program employ a user driven approach and recognize the substantial investments that public safety and homeland security agencies have already made in existing equipment and procedures. Additionally, this office must recognize the challenges associated with incorporating legacy equipment and practices in the face of constantly changing technology. Driving principles behind the management of this office include:

1. *Recognizing that it must be a user driven program*—Emergency response providers and homeland security practitioners—who own, operate and maintain more than 90% of the nation's wireless public safety infrastructure—will be integrated into the program from its beginning, thereby allowing the program to create solutions that meet their needs. The public safety community will be involved primarily through associations. There are two reasons for this approach: (1) the associations represent the leadership of their respective constituencies; and (2) as the National Task Force on Interoperability (NTFI) has demonstrated, the associations are an excellent way to reach out to these communities.
2. *Extensive leveraging of what exists*—Cooperation and coordination with existing programs reduces unnecessary duplication of effort and increases efficient use of Federal resources dedicated to common causes. In addition, the investments that many public safety agencies have already made must be maximized.
3. *A standards-based approach*—Standards maximize competition across industry, encourage technology innovation, create an overall cost savings, and increase compatibility among public safety and homeland security agencies.
4. *Compliance with key policy documents and initiatives*—Compliance with the National Incident Management System, the National Response Plan, and relevant Homeland Security Presidential Directives will provide a consistent nationwide approach for agencies at all levels of government to work effectively and efficiently together to prepare for, prevent, respond to, and recover from major incidents.
5. *An effective outreach program*—Outreach efforts will emphasize the need for interoperability, and tools for its implementation, to practitioners and policy makers at all levels of government, and the public safety community.

Portfolio Example: Communications Interoperability

As a public safety practitioner driven program, and as part of OIC, SAFECOM is working with existing Federal communications initiatives and key public safety stakeholders to address the need to develop better technologies and processes for the cross-jurisdictional and cross-disciplinary coordination of existing systems and future networks. SAFECOM has three objectives: (1) developing standards in partnership with Federal, State, local, and tribal public safety organizations to define the requirements for first responder interoperability at all levels; (2) building from those standards, developing a national architecture in coordination with the work under the National Response Plan to assist in the progression towards wireless interoperability; and (3) developing and implementing a process to coordinate the Federal government's wireless interoperability investments and programs. The customer base includes over 50,000 local and State public safety agencies and organizations. Federal customers include over 100 agencies engaged in public safety disciplines such as law enforcement, firefighting, public health, and disaster recovery. Because it is a government-wide E-Gov initiative, SAFECOM is not a part of the S&T's FY 2005 budget request. Rather, SAFECOM is currently funded by multiple partner agencies that transfer funds to DHS.

SAFECOM Achievements To Date

Over the last year, SAFECOM has made significant progress in both achieving its short-term goals and building the foundation for a longer term, comprehensive program. It has established itself as the umbrella program within the Federal government coordinating with local, tribal, State, and Federal public safety agencies to improve public safety communication and interoperability.

• *Coordinated Funding Assistance*—In FY 2003, SAFECOM developed grant guidance in keeping with the needs of public safety for use by Federal programs funding public safety communications equipment to State and local agencies. Community Oriented Policing Services (COPS), Federal Emergency Management Agency (FEMA), and ODP incorporated this guidance into their public safety communications grants. This guidance marked the first coordinated approach to funding requirements. In further support of the coordinated grant process, SAFECOM organized and funded the peer review process for the joint grant solicitation from COPS and FEMA. SAFECOM also supported the Department of Commerce National Institute of Standards and Technology (NIST) Summit on Interoperability that was the first step towards identifying all the Federal and national programs involved in public safety communications so that a broader coordination effort can continue.

- *Statement of Requirements Development*—SAFECOM recently developed the Statement of Requirements (SoR) for Wireless Public Safety Communications and Interoperability in coordination with the National Public Safety Telecommunications Council, NIST, and the Department of Justice's AGILE Program. The SoR contains interoperability scenarios describing how SAFECOM envisions technology enhancing public safety. From these scenarios, operational requirements are defined and functional requirements of the technologies are extrapolated. The requirements identified in the SoR will drive the development and creation of interface standards that will satisfy public safety practitioner needs. The SoR will also offer industry a resource for understanding the users' needs in the development of new technologies and serve as a guide for SAFECOM to develop its research development, test, and evaluation program and constitutes the first national definition of what interoperability must accomplish.

SAFECOM is on track to achieve these critical milestones in 2004:

June: SAFECOM Strategic Plan Update

- SAFECOM will conduct a strategic planning session at the Executive and Advisory (EC/AC) Committee Meetings in June. The EC and AC are comprised of senior level stakeholders from the local, State, and Federal public safety communications communities. At this time, strategic initiatives developed at the December Joint Planning Meeting will be reviewed, and new objectives will be developed for the short and long term goals of the program. Afterwards, SAFECOM will produce and distribute a modified strategic plan based off the stakeholder comments presented at these meetings.

July: Detailed Interoperability Project Plan for Virginia

- SAFECOM will develop a detailed project plan using the result of the strategic planning session and the project team's technical expertise. This project plan will include tasks that need to be accomplished by the Commonwealth along with realistic timeframes for completion. Like the Virginia Strategic Planning Session, this plan will serve as a model for other States as they work towards achieving communications interoperability for public safety first responders.

August: Interoperability Grant Peer Review

- SAFECOM will facilitate interoperability grant peer review sessions enabling public safety communications subject matter experts to evaluate and comment upon grant applications for FY 2004 COPS and FEMA communications equipment grants. These reviewers will ensure that grants will be distributed only for projects that meet SAFECOM developed interoperability requirements.

September: RapidCom9/30 Completed

- SAFECOM is undertaking an initiative to ensure a minimum level of public safety interoperability is in place in ten key urban areas by September 30, 2004. The RapidCom9/30 project will provide incident commanders in charge of managing/directing various responding agencies the ability to adequately communicate with each other and the respective command center within one hour of an incident. Due to this effort's limited scalability, it is not meant to serve as comprehensive public safety communications solutions, but as an interim solution that provides minimum interoperability capability during emergency responses.

September: Narrowbanding Report Released

- SAFECOM will release a report detailing the program's recommendations on spectrum policy in regard to narrowbanding in the 700 MHz band. As recent events in the 800 MHz band have shown, coordinated spectrum policy is important for public safety communications, and SAFECOM's input to any plan in the 700 MHz band will allow for more efficient spectrum use when allocated frequencies become available in the next decade.

September: National Guard Study Released

- SAFECOM will release a report outlining how National Guard Land Mobile Radio (LMR) resources can be incorporated into the plan to achieve nationwide interoperability. It will also identify how local public safety organizations can leverage National Guard assets. The National Guard already has a great deal of investment in LMR facilities, and this report will help local and State public safety organizations utilize resources that may already be present in their communities.

October: Communication Device Report Released
- SAFECOM will release a report detailing the findings of its testing and evaluation program. The first report will focus on the performance of public safety communications equipment with the P25 Phase I standard. This report is the first step in developing a comprehensive national architecture plan for communications interoperability.

November: Portal for Interoperability Information goes live
- The Web Portal of Interoperability Information will be the "One-Stop-Shop" for information pertaining to public safety communications interoperability. As a portal, it will be an interactive community space, allowing registered users to research potential solutions as well as share their thoughts on existing technologies. Version 1.0 of this portal, which will be released in November, is the first attempt to provide first responders with a central repository of critical information pertaining to communications interoperability.

December: National Interoperability Baseline Methodology Released
- SAFECOM will release a methodology detailing how a baseline of the level of interoperability nationwide can be established. The baseline is required in order to understand the current level of interoperability at the local and State levels and will be used to measure the success of the SAFECOM Program in achieving national communications interoperability for first responders in the coming years.

Conclusion

Our nation is heavily invested in an existing infrastructure that is largely incompatible. As I stated earlier, current efforts within the Federal government to address the interoperability problem are being coordinated to incorporate the needs of local, State, and Federal practitioners. We must continue to pursue the current comprehensive strategy that takes into account technical and cultural issues associated with improving communications and interoperability, and recognizes the challenges associated with incorporating legacy equipment and practices given the constantly changing nature of technology.

The many obstacles facing public safety interoperability and compatibility make for complex problems with no one-size-fits-all solution. Flexible and dynamic resolutions are necessary to combat the unique challenges presented by distinct localities and States. The new OIC, with its partners, will work towards a world where lives and property are never lost unnecessarily because public safety agencies are unable to communicate or lack compatible equipment and training resources.

Mr. UPTON. Thank you.

Mr. Grube?

STATEMENT OF GARY GRUBE

Mr. GRUBE. Good afternoon, Chairman Upton, Ranking Member Markey and members of the subcommittee. My name is Gary Grube, and I am the chief technology officer of Motorola's Public Safety Communications Enterprise, and I have worked with the first responder community for nearly 25 years.

I want to thank you, Mr. Chairman, for scheduling this hearing, for your committed leadership on communications matters and for focusing on the needs of the Nation's first responders. It is an honor to be here with you today to discuss mission critical, interoperability communications capabilities.

Chairman Barton and Ranking Member Dingell, thank you for the excellent work earlier this month in marking up the DHS authorization bill. It now addresses the need for the deployment of communications equipment based on national voluntary consensus standards. As you know, a standard called Project 25 is the open standard that has been endorsed by every major law enforcement organization in the country.

This hearing follows quite nicely the one you held on June 2nd on the FCC's digital television transition plan. Motorola is highly encouraged by this initiative. The committee leadership sent a real message that change is afoot. Chairman Barton's leadership and proposal changed the terms of the debate. We would like to express our deep appreciation for the positive new direction he is setting for the transition.

I also want to thank the other Congressmen Fossella, Stupak, and Engel who have been exploring ways to help get the funding first responders need. Motorola has been a leading provider of public safety solutions for over 65 years. Wireless communications provide our first responders with the right information, at the right time and in the right place, whether that information is voice, data or video.

Today, the technology exists to improve the quality and effectiveness of public safety operations, but there are two obstacles to deploying these new technologies. First, public safety must have access to the 700 megahertz spectrum by year-end 2006 to deploy interoperable voice and advanced data technology as early as possible. This spectrum can literally save lives.

Second, public safety needs additional Federal funding to purchase the radios and systems necessary to do its job. When these steps are taken, advance wireless technology can fully support our first responders. Together we can improve the quality of mission-critical information to our front-line responders.

An officer or agent could transmit video of a potential bomb or biological weapon and get real-time counsel from an expert in another location. Local or State police could instantly send or receive a photograph of a missing or abducted child. Firefighters can access building blueprints, hydrant locations, hazardous material data and other critical information.

We have heard a great deal about the need for improved interoperability among first responders organizations. Some Federal funds have been made available for this purpose, but they are inadequate to reach an acceptable level of interoperability in a reasonable time. We need congressional leadership committed and enforcing a sustained well-funded, multiyear Federal program that guarantees this communications problem will be fixed.

Turning to the need for spectrum. In 1997, this committee and the FCC recognized its importance by allocating spectrum in the 700 megahertz band for mission-critical State and local public safety communications. This spectrum continues to be used for TV and needs to be cleared.

This spectrum is critical to public safety operations for two reasons. No. 1, 700 megahertz provides additional capacity for interoperability and voice communications. And, number 2, 700 megahertz is the only dedicated spectrum allocation where public safety can have high-speed data, wide area access in the field to data bases, the Internet, imaging and video, or, in other words, critical information.

Unfortunately, most metropolitan area public safety operations cannot use the spectrum today, nor can they predict with any certainty when they might have access to these frequencies. This uncertainty is due to the way the current law is written. In reality, there is no hard date for ending the DTV transition, leaving public safety and deployment of vital technology in limbo. Until this problem is addressed, 5 percent of this Nation's TV stations block improved public safety communications for over 50 percent of the population. We are mindful of the other considerations that are involved in clearing these channels, and we believe that the adverse effects can be mitigated.

At a hearing last year, this committee asked about the impact on TV viewers. Using independent data, we have determined that, on average, only 3 percent of the TV households covered by these blocking stations actually tune in over the air during a typical week. As we explore ways to resolve the transition, we encourage you to continue your examination of the Berlin model which delivered a crisp analog cutoff date using digital-to-analog converter boxes. This ensured a seamless changeover for all TV consumers.

Motorola is completing its analysis, and we expect to place on record at the FCC an estimate in the sub-$100 range per unit for a digital-to-analog converter that would inexpensively facilitate a Berlin model type solution in the U.S.

Even more spectrum may be required in the band to support homeland security coordination among Federal, State and local agencies and critical infrastructure entities. For example, a wide area broadband pilot here in the Capital demonstrates the need for such additional spectrum.

In closing, Mr. Chairman, making the public safety spectrum available nationwide by the start of 2007 will not happen without your commitment and your help. The first step is to agree today to set that hard date. We urge this committee to clear the spectrum and to invest in interoperability for all public safety radio users. Motorola pledges its support to our customers and to this committee to make this happen. Thank you.

[The prepared statement of Gary Grube follows:]

PREPARED STATEMENT OF GARY GRUBE, CORPORATE VICE PRESIDENT AND CHIEF TECHNOLOGY OFFICER, COMMERCIAL GOVERNMENT AND INDUSTRIAL SOLUTIONS SECTOR, MOTOROLA

Good afternoon, Chairman Upton, Ranking Member Markey and Members of the Subcommittee.

My name is Gary Grube, and I am the Chief Technology Officer of Motorola's business sector that serves state and local public safety and Federal law enforcement customers. I have worked with the 1st responder community for nearly 25 years.

I want to express my appreciation to you, Mr. Chairman, for scheduling this hearing to address such an important issue as improving interoperability for our nation's Police, Firefighters, Emergency Medical Personnel and Federal agents. It is an honor to be here with you today to discuss mission critical interoperable communications capabilities.

I would be remiss if I did not thank you, Chairman Barton, and Ranking Member Dingell for the excellent work you undertook earlier this month in marking up the Select Committee on Homeland Security's DHS authorization bill. It now addresses the need for the deployment of communications equipment based on national voluntary consensus standards. As you know, a standard called "Project 25" is the open standard that has been endorsed by every major law enforcement organization in the country. And, because it delivers true interoperability, the FCC has set P25 as the interoperability standard in the 700 MHz band.

This hearing also follows quite nicely the one you held on June 2nd on the DTV transition. You heard testimony from Mr. Ferree on the FCC's Media Bureau Plan to advance the DTV transition and thereby provide needed spectrum to public safety by the start of 2009. Motorola is highly encouraged by this initiative. At that hearing, the Committee leadership sent a real message that change is afoot and that the American public and their heroes can look forward to date certain availability for spectrum for critical interoperable communications. Chairman Barton's powerful words and proposal resonated with us and the public safety community, and we'd like to express our deep appreciation for the positive direction he is setting for this debate.

I also want to thank the other Members of this Committee, notably Congressmen Fossella, Stupak, and Engel who have been exploring ways to usher in new 1st responder high-speed communications and to find additional funding mechanisms to enable them.

Meeting these communications needs is critical to the safety and well being of our first responders and the entire American public they serve. I am pleased to be with you today to support your efforts to achieve our shared goal of meeting public safety's communications needs.

I'd also like to note that it is good to be at the witness table with David Boyd, who heads the SAFECOM program at the DHS. Mr. Boyd works very closely with State and local 1st responders and is very supportive of their interoperable communications equipment needs.

Motorola is a leading provider of communications and information solutions, with more than 65 years of experience in meeting the mission critical needs of our public safety customers. We offer an extensive portfolio of solutions specifically designed to meet the rapidly evolving safety and security needs of these customers. Our solutions include interoperable mission-critical radio systems based on the P25 public safety interoperability standard; command and control solutions; identification and tracking solutions; information management for criminal justice and civil needs; and physical security and monitoring solutions.

In 2002, my business sector in Motorola received the Malcolm Baldrige National Quality Award, the nation's premier award for performance excellence and quality achievement. We continually strive to translate the quality processes upon which this award was based into high quality and reliable communications systems for our public safety customers. Motorola works very closely with our customers to help them implement communications capabilities needed for both every day mission critical needs and catastrophic events.

PUBLIC SAFETY REQUIRES DEDICATED MISSION CRITICAL SYSTEMS

Our partnership with the public safety community over the years has taught us that first responders need systems designed specifically for mission critical operations to get the job done. As with most of the Northeast and Midwest, the State of Michigan was confronted with a large-scale emergency during the August 2003 blackout. Despite the failures experienced by various commercial carrier networks in Michigan and surrounding states due to these power outages, Michigan's nearly 12,000 public safety radios experienced no interruptions in communications. Police officers, firefighters and EMS providers worked as a team in real time to serve the public. Michigan had control over its communications because it had created a statewide critical network designed specifically for catastrophic situations and events, including the disruption of normal power sources. While many public safety entities also use public carrier networks for less critical communications, there is no substitute for mission critical systems when the safety of first responders and the public they serve is at risk.

TRUE INTEROPERABILITY REQUIRES A SUSTAINED FOCUS

Ask any firefighter, police officer or EMS provider and they will tell you that the ability to communicate reliably, instantly and securely is one of the most critical factors in managing a crisis situation. For almost all first responders, a handheld radio device is their communications lifeline—giving them the ability to communicate during a crisis situation. While the most visible part of the communications system to first responders and the public, these handheld devices must be supported by communications network infrastructure. Together the system of infrastructure and radios must be designed to provide the necessary coverage, capacity, reliability and features required for mission critical operations. Yet, despite the Federal prioritization of homeland security, a large number of first responder radio systems are not yet truly interoperable and simply cannot talk to each other in a crisis situation. While public safety agencies are making progress on improving communications capabilities and interoperability, much more remains to be done. This problem will not be solved overnight. There is no "quick fix" solution for true interoperability. Providing true interoperability for the nation's first responders will require a multi-year dedication and focus on the part of Congress, the public safety community and industry.

There are four key foundation blocks to achieving improved public safety communications capabilities and interoperability. These are 1) sufficient spectrum, 2) adequate funding, 3) use of standardized mission critical technology, and 4) operational planning and practice.

I'll address these briefly, and then in more detail.

Spectrum that could significantly improve interoperability of public safety communications has been allocated but is not yet accessible in most major markets. Additional spectrum allocations are also needed. The Administration and the Congress have begun to fund the various grant programs administered by the Departments of Justice and Homeland Security and to set interoperability as a high priority for these funds. However, the level of funding in general and the amounts set aside for interoperable equipment purchases must be increased significantly and sustained over multiple years to deliver on this goal.

Interoperability standards that meet public safety needs and are open to all manufacturers have been established for voice and data communication and for wideband services. A broadband standards development initiative is also underway. Communications technology meeting the Project 25 (P25) voice and data interoperability standard developed by the public safety community and industry is available from multiple equipment vendors. Wideband and broadband technologies capable of meeting public safety's increasing need for high speed data and imaging have been developed and are being trialed.

Finally, Pubic safety users realize now, more than at any time in history, the value of planning and practice among multiple agencies, jurisdictions, and levels of government.

The remainder of my testimony addresses in more detail the four foundation blocks and what Congress can do to help public safety improve communications capabilities and interoperability.

REAL ACCESS TO MORE PUBLIC SAFETY SPECTRUM IS ESSENTIAL.

As discussed above, effective mission critical mobile and portable communications systems are absolutely essential to public safety operations. Police officers, firefighters, emergency medical personnel and their departments use mobile and portable communications to exchange information that can help protect public safety officials and the citizens they serve. Traditionally, this information was mostly exchanged by voice. Increasingly, as public safety entities strive to increase efficiency and effectiveness in today's world, they also need the capability to transmit and receive high performance data, still images and video reliably. Spectrum is the road upon which such communications travel, and increased communications requirements lead to the need for more spectrum.

Based on a thorough justification of need, Congress and the Federal Communications Commission dedicated 24 MHz of spectrum in the 700 MHz band to State and local public safety in 1997. The FCC established specific nationwide interoperability channels within this spectrum allocation, as well as both narrowband and broadband channels to support a variety of identified public safety communications requirements. However, seven years later, incumbent television stations operating on channels 62, 63, 64, 65, 67, 68 and 69 prevent public safety access to this essential resource in most major urban areas where the demand for more spectrum is the greatest. The recent focus on increased interoperability and Homeland Security make availability of this public safety spectrum nationwide even more critical.

These channels are critical to public safety for two reasons:

(1) Together, the new 700 MHz and current 800 MHz bands provide the best opportunity to integrate interoperable communications. The 700 MHz band's proximity to the 800 MHz band allows public safety agencies to expand their current 800 MHz narrowband voice and data systems for interoperability and regional coordination on an "intra" as well as "inter" agency basis. Equipment operating in these combined frequency bands on the FCC endorsed Project 25 interoperability standard is commercially available today. The FCC has granted each state a license to operate such narrowband communications in the 700 MHz band. Localities throughout the country are actively engaged in spectrum planning at 700 MHz, a prerequisite for obtaining their own FCC licenses. For example, after a yearlong review by the FCC, the Southern California regional plan was recently approved, but TV incumbency prevents actual use of the spectrum in much of that area.

(2) 700 MHz is the only dedicated spectrum allocation where public safety can implement advanced mobile wide area systems that bring high-speed access to databases, the intranet, imaging and video to first responders out in the field.

This technology offers a whole new level of mobile communications capabilities, which is far beyond today's voice and low speed data applications. For example:

a. An officer or agent could transmit video of a potential bomb, or biological weapon and get real time counsel from an expert in another location.

b. Local or state police could instantly send or receive a photograph of a missing or abducted child.

c. Crime scene investigators can transmit live video of footprints, fingerprints and evidence to speed analysis and apprehension of perpetrators.

d. Firefighters can access building blueprints, hydrant locations hazardous material data and other critical information.

e. Paramedics can transmit live video of the patient to doctors at the hospital that would help save lives.

Motorola previously conducted wideband trials together with public safety entities in Pinellas County, Florida and the City of Chicago, and we are currently participating in the District of Columbia's broadband trial. As to the Chicago trial, we greatly appreciate Chairman Upton leading a delegation of Committee Members, including Congressmen Bass, Rush, and Terry to participate in a demonstration last year with the Chicago Police Department. We would like to encourage a similar delegation to see the outstanding broadband trial that is being led by Robert LeGrande on behalf of the DC Government. We are proud to be working with him on an innovative solution that will deliver powerful applications to the frontline 1st preventers here in our Nation's Capitol. All of these trials operate under experimental 700 MHz licenses from the FCC. The capabilities demonstrated are the emerging powerful multi-media applications that will bring public safety communications into the Twenty-First Century.

Public safety users and industry finalized the wideband interoperability standard, TIA902, through the Telecommunications Industry Association (TIA). Public safety has recommended that standard to the FCC for the 700 MHz wideband channels, and we are anxiously awaiting FCC action on that request. Right now, actual product development could proceed as soon as we know with certainty that this spectrum will be available nationwide to the public safety community.

Unfortunately, most metropolitan area public safety operations cannot use this spectrum today, nor can they predict with any certainty when they might have access to these frequencies because of incumbent TV operations. Therefore, public safety users in most cities cannot deploy, or firm up plans for the actual deployment of, improved interoperability and advanced capabilities that will improve their effectiveness and safety.

Current law and policies set December 31, 2006 as the date for clearing television from the band. However, this is not a firm date. Broadcasters do not have to clear the band until 85% of the households in their service areas have the capability to receive digital TV, an environment unlikely to be met in most markets by yearend 2006 under the current rules. Under current law, while TV incumbents are required to vacate this spectrum at the end of 2006, they can receive an *unlimited* extension of this deadline based on the state of the transition in their particular market. So, in reality, there is no "hard date" when the transition will end and the spectrum will really be accessible to public safety everywhere. This is not the optimal situation for the public safety community and those they serve. We commend and encourage efforts by this Subcommittee and the FCC to ensure that this spectrum is cleared nationwide for public safety use no later than yearend 2006.

The reality is that 5% of this country's TV stations are blocking improved public safety communications for 84% of the population in the largest cities, those over 200,000. Of that 84%, more than two-thirds have no access to the spectrum, while the remaining third have only limited access. When we look at all areas of the country, rural as well as urban, 54% of our country's population is totally blocked by this relatively small number of TV stations from receiving any benefits of public safety communications in this band.

In a hearing before this Subcommittee in June, 2003, Greg Brown of Motorola testified about the need for access to the 700 MHz spectrum. During that hearing, Subcommittee Members acknowledged this need, but also discussed the potential impact on some TV operations of setting a firm date for broadcasters to finally return their analog TV channels in the 700 MHz band.

The concerns expressed at that hearing spurred us to perform a study to determine the impact on the viewing public of clearing that spectrum. That study "700 MHz TV Clearing and its Impact on TV Viewership" is attached in its entirety. As shown in this study, the potential harm to the viewing public is limited. And the benefit to public safety is dramatic.

First, only 75 stations, equaling less than 5% of the more than 1500 U.S. TV stations, affect public safety's availability of its Congressionally mandated 700 MHz band frequencies. Second, Motorola's analysis of independent television industry data shows that, on average, only 14% of the TV households who have the option to view these stations actually do so at all, and that of those viewing, 82% watch by cable. *This means that, on average, only 3% of the TV households within these stations' coverage areas actually tune to these stations over-the-air sometime during an average week.*

The Committee is also aware of an FCC plan that would complete the analog to digital TV transition by January 1, 2009. We applaud the FCC for taking the leadership and initiative to move the debate toward a successful conclusion. While 2009 may be an appropriate date by which all 1500 or more TV stations would complete the transition, the public safety community has stated that its needs justify clearing the 5% of stations blocking its 700 MHz band channels by 2007. By yearend 2006, public safety will have waited almost ten years to access this spectrum.

As noted above, very few TV households would notice any significant impact of clearing this spectrum for public safety. Those that do could be provided with an inexpensive digital-to-analog over-the-air converter box. Motorola is a TV set-top box provider. That business unit is presently completing its analysis, and we expect to place on the record at the FCC a sub-$100 estimate per unit for an over-the-air digital-to-analog converter that would help to facilitate a Berlin Model-type solution in the US. We understand the Committee and the GAO are already reviewing the actions undertaken in Berlin, Germany to ensure a seamless and pain-free crisp analog to digital TV transition. This was achieved through the provision of converter boxes to some TV consumers who did not subscribe to cable or satellite service and maintained an analog TV set. We believe this is a positive step that could provide a real path forward on how to solve the transition here in the U.S.

Congressional action is required to ensure that TV incumbents return this critically needed spectrum, without exceptions, by a firm date—which should be no later than yearend 2006.

We urge the Committee not to be deterred from setting this goal because it has been hard to achieve to date. Rather, once it has been set, the affected parties, including the public safety community, the FCC and NTIA, the involved broadcasters and other affected parties, including our company, should be called upon to devote our energies to making it happen.

As you know, the 24 MHz of spectrum in the 700 MHz band is allocated for State and local public safety use. That spectrum, if cleared, would only partially satisfy the spectrum need documented by the public safety community. No comparable spectrum exists for meeting the Homeland Security requirements of Federal agencies or critical infrastructure entities. Such interoperability among State and local first responders, Federal agencies and critical infrastructure entities will best be achieved through the availability of comparable spectrum resources. Therefore, we recommend that Congress consider meeting these additional needs by reallocating the remaining 30 MHz of commercial spectrum in the 747-762 MHz and 777-792 MHz portions of the band which are presently targeted for auction. This spectrum should be reallocated as a Homeland Security band to support State, local, Federal and critical infrastructure (such a utilities and nuclear facilities) communications needs.

We also note that a spectrum coalition headed by Mr. LeGrande, in the District of Columbia Office of the Chief Technology Officer (OCTO), has requested that 10 MHz of additional spectrum at 700 MHz be designated for broadband use. Since that 10 MHz falls within the 30 MHz recommended for reallocation here, we believe that request and reallocation of the 30 MHz are complementary to one another. Motorola is quite pleased to be one of the partners with OCTO in trialling 700 MHz broadband systems and public safety applications.

As part of this reallocation, Congress should charter a committee of key representatives from major public safety associations, Federal agencies and critical infrastructure entities to determine how that additional 30 MHz of spectrum should be distributed among State, local, Federal and critical infrastructure entities.

Should the government wish to pursue this important reallocation of spectrum, anticipated auction revenue from these 30 MHz of spectrum would no longer be available. However, we believe substitute spectrum that could provide potentially stronger auction receipts can be identified to replace this anticipated revenue and could be used to support a Berlin Model-type subsidy solution domestically. Motorola greatly appreciates this Committee's continued policy thrust to find ways to reinvest spectrum auction revenues in ways to advance technology deployment and economic development, whether it is the Commercial Spectrum Enhancement Act that this body passed last year and is under active consideration in the Senate, or the Chairman's proposal to use auction revenue to help support the return of the analog TV frequencies for other valuable services—including interoperability.

PROJECT 25 IS THE U.S. INTEROPERABILITY STANDARD FOR MISSION CRITICAL OPERATIONS

In addition to spectrum access, standardized technology is critically important to achieving interoperability. Fortunately, the public safety community and multiple manufacturers have partnered to develop a suite of standards for interoperability known as Project 25.

Public safety users adopted the P25 standard in order to implement an open standard that promotes interoperability and system migration, and enables more competitive procurements for digital radio systems and radios—thereby eliminating dependence on one vendor for radios, even after their systems have been installed.

P25 is actually a full suite of standards that, when built into communications equipment, provides the basis for interoperable digital radio voice and low-speed data communications among multiple public safety users, departments and agencies. These standards were developed under the auspices of, and are published by, the Telecommunications Industry Association (TIA), and accredited by the American National Standards Institute (ANSI). Public safety users led the development of the standard and have the option to choose Project 25 products from multiple vendors.

Unlike many other communications standards and technologies in the broader wireless industry, the unique mission critical requirements of public safety users drove the development of the P25 suite of standards. High priority was given to public safety's operational and tactical requirements. For reasons of cost effectiveness, the Project 25 standards permit a graceful migration path from aging analog to new digital systems. These standards promote improved spectral efficiency, and, as intended, allow for multi-vendor equipment offerings. Radios that meet the P25 standards incorporate backward compatibility with conventional analog systems. Project 25 radios communicate in analog mode to analog radios, and either digital or analog modes with other P25 radios.

Public safety users at all levels of government have embraced Project 25. For example, P25 has received the endorsement of the National Association of State Telecommunications Directors (NASTD), the Association of Public Safety Communications Officials—International (APCO), the International Association of Chiefs of Police (IACP), the International Association of Fire Chiefs (IAFC), the Major Cities Chiefs (MCC), the National Sheriffs' Association (NSA), and the Major County Sheriffs' Association (MCSA).

Project 25 has received broad support at the Federal level as well. Based on public safety user recommendations, the FCC endorsed the Project 25 suite of standards for voice and low-speed data interoperability in the new nation-wide 700 MHz frequency band. Every 700 MHz radio must include Project 25 compatibility defined by this TIA/ANSI standard, and the FCC set P25 as the required mode of operation on the 700 MHz interoperability channels. The U.S. Department of Defense mandated P25 for new land mobile radio systems. The Department of Homeland Security has also endorsed P25 as the preferred standard for digitally trunked radio systems as part of its Federal grant guidance.

INTEROPERABILITY FUNDING SHOULD BE A NATIONAL PRIORITY

Full public safety communications interoperability within the decade should be a national goal. This is an ambitious goal, but a very worthy and doable one. Our nation has the necessary technology, the standards and equipment. After spectrum, what is lacking are the economic resources to acquire the equipment and deploy the systems, particularly at the state and local level, and we will not achieve this goal at the present pace of system upgrades. Instead, it will require a commitment lead by determined champions. Mr. Chairman, I urge this Committee to assume this important role.

There are several reasons why the Federal government must take the lead. As we all know, homeland security is a Federal, State and local responsibility, but national planning begins at the Federal level. This is one of the reasons why the Congress and the President created the new Department of Homeland Security.

While we cannot predict future terrorist attacks, we must prepare for the real possibility and threat. Also, we do know that we will face natural disasters such as hurricanes, tornados, wildfires, and earthquakes and other threats such as hostage takings, hazardous materials spills, and train wrecks. Interoperable public safety communications are critical to effective response in all these cases.

The states face a staggering $80 billion aggregated deficit in FY2004 alone, and this puts serious limits on their spending. As a result, they cannot be expected to accomplish this goal without substantial Federal support. Accordingly, we need a well-funded, multi-year Federal program that guarantees that this communication problem is fixed, once and for all.

Consequently, we must work aggressively to increase the funds devoted to interoperable communications now and until the job is done. Nothing should be allowed to delay or impede this funding effort. In FY 2004, approximately $4.4 billion was appropriated for Federal equipment grant programs for State and local first responders.— However, wireless communications is only one of a large number of allowed uses for these funds. Only about $85 million or 2% of the total was designated in the legislation specifically for wireless communications enhancements.— We would ask for your help to increase the sums designated for wireless communications in light of the broad consensus that exists for improving the status of wireless communications interoperability among government entities. If we are going to fix the interoperability problem we must have a well-defined goal, a program to achieve that goal, and a way of measuring programs that is visible to the Congress.

We certainly cannot afford the human costs associated with delaying achievement of full interoperability.

PLANNING AND PRACTICE ARE ALSO ESSENTIAL FOR INTEROPERABILITY

Planning for interoperability at the operational level is also a key element of improving interoperability. In situations where multiple agencies and jurisdictions have planned operational procedures and practiced that plan, interoperability has improved. For example, multiple agencies can decide in advance how best to organize communications groups to support the various responders at an incident scene. Practice drills help public safety responders become familiar with these procedures so they can be more easily implemented at an actual emergency incident.

Planning and practice are supplements to, not substitutes for, adequate spectrum, funding and technology. All elements of the foundation need to be in place to improve public safety mission critical interoperability and capability. While Congress has the greatest influence over the interoperability building blocks of spectrum and funding, public safety agencies are the focal point for planning and practice.

Mr. Chairman, ensuring that our nation's public safety officials have the tools they need to protect our citizens in the years ahead is a sound investment for the entire country. We urge this Committee to clear spectrum for public safety and to champion investments in interoperability for all public safety radio users. Motorola pledges its support to our public safety customers and to this Committee to help you make this happen.

Thank you.

Mr. UPTON. Thank you.

Mr. LeGrande?

STATEMENT OF ROBERT LEGRANDE

Mr. LeGRANDE. Good afternoon, Mr. Chairman and members of the subcommittee. My name is Robert LeGrande, and I am a deputy chief technology officer for the Office the Chief Technology Officer, the central information technology and telecommunications agency of the District of Columbia government. I am responsible for wireless communications infrastructure for the District government and a representative of the Spectrum Coalition for Public Safety.

Over the past year and a half, I have led wireless public safety voice and data communications programs for the District of Columbia. In this role, I have partnered with executives, communication decisionmakers and field personnel of the Metropolitan Police Department and the Fire and EMS Department to upgrade our public safety voice network and install public safety broadband wireless networks.

During this process, I gained tremendous respect for the work of our first responders and gained an even greater appreciation of their communications needs. Today, I will describe for you the efforts and the results of the voice communication upgrades, which include local, regional and Federal interoperability. I will also describe the Spectrum Coalition for Public Safety's efforts to secure additional megahertz of 700 spectrum which will enable Public Safety to build and deploy broadband wireless networks throughout the United States.

Please reference the diagram with the city configuration here to my right. This diagram represents both the accomplishments of a wireless voice and the vision of our wireless broadband communications within the District. As depicted in the diagram, our recently upgraded 10-site radio network provides comprehensive in-building coverage, augmented by 63 vehicle repeater systems to provide the highest level of coverage available to first responders.

In addition, by using distributive antenna techniques, we provided for the first time nearly 100 percent coverage in the underground subway system. It is important to emphasize that interoperability is accomplished individual by individual, and I say this because we must first ensure that our first responders can communicate clearly in all areas of the city before we focus on communications outside or with other agencies. Without sufficient radio coverage, intraoperability, much less interoperability, is impossible, putting lives at risk, even for day-to-day first responder events.

This wireless infrastructure will soon ride on the city's fiber optic network. DC Net delivers the highest level of redundancy and reliability for our first responders.

In order to achieve interoperability, we took several steps. First we had to upgrade the coverage and capacity or our preexisting non-interoperable local networks. We accomplished this by creating a single dual-band radio network. Next, we had to create interoperability on our intra-District public communication systems and other first responders in the region.

Please reference the Rubik's Cube depiction of the DC-based regional public safety interoperability diagram. When my team first shared this diagram with me, I simply hated it, because it was too complex and too hard to understand. Later, my thought became, "Exactly." Interoperability among many jurisdictions is very complex and hard to understand. In this diagram, we only note the interoperability methodologies for the District of Columbia. Please understand, every agency that is on the left of that diagram must have a similar Rubik's Cube representing its interoperability methodologies. All of these puzzles must be figured out for all the agencies represented in order to achieve interoperability in the region. It is a very complex process.

I am very pleased to report to this committee that we have made substantial improvements in public safety voice communications, and a detailed status of the DC interoperability progress is provided in attachment 3 of my testimony. These improvements would not have been possible without Federal investment and the coordinated efforts of our Regional Council of Governance, which is made up of first responders, which is made up of first responders, leaders and communications specialists from our surrounding region.

Additionally, and without them prompting me to say this, without clear, unambiguous direction from our congressional leaders, our mayor, city council, city administrator, deputy mayor for public safety, police and fire chiefs and my boss, Suzanne Peck, the chief technology officer of this city, we would have simply failed that direction: Work together, get it done now.

The interoperability effort considered several options. We chose to create dual band 460 and 800 megahertz network because we could use Motorola's existing SmartZone architecture to incorporate all of the existing 800 megahertz radios and over half of the 460 megahertz radios. This allowed neighboring agencies using 800 megahertz radios to communicate with our 460 megahertz police radios through a central hub, and through trunk radio technology delivered the maximum number of individual communication paths for numerous simultaneous incidents and talk groups.

As mentioned previously, our underground project, managed jointly with WMATA and completed in March of this year, remedied coverage problems for the District's fire and EMS first responders in the WMATA subway system, upgrading the 800 megahertz underground distributive antenna system. District firefighters now have seamless coverage for above and below ground, and they can actually ride the train and achieve a high level of voice quality. Together, these two projects gave the District one of the best public safety wireless voice systems in the Nation: Comprehensive coverage, 27 channels, a regionally interoperable system providing clear voice communication, encryption and other digital features.

During the requirements and design phase of our voice programs, we realized that providing upgraded voice communications for first responders is simply not enough. The threats to our country and region are real an imminent. Providing our first responders with city-wide remote surveillance, chemical and biological and bomb detection systems is critical to preventing attacks.

Additionally, early detection of attacks will speed our response capabilities. We evaluated the use of commercially available wireless networks, wideband wireless networks and networks deployed at the 4.9 gigahertz spectrum, and none of these met our requirements. Please note, individuals and organizations that wish to do our country harm already have city-wide broadband wireless capabilities in the District of Columbia, North Carolina and San Diego. They can sign up anonymously for Verizon or Nextel's services in these areas and conduct real-time broadband intelligence gathering and video surveillance; worse, attack coordination for far better coordination capabilities than was used in Madrid, Spain. Our first responders need better tools than the terrorists.

Recently, the District launched initiatives aimed at delivering the next generation broadband wireless solutions in the Nation's Capital, and potentially the Nation. We founded the Spectrum Coalition for Public Safety for 30 states, counties, cities, regions and public safety organizations. The goals of this coalition is to pursue legislation that require the FCC to reserve an additional 10 megahertz of radio spectrum for wide area public safety broadband wireless uses, enabling competitive, affordable technologies that meet first responder requirements and facilitate nationwide network deployment.

Concurrently, the District is deploying on a pilot basis the Nation's first citywide wireless broadband public safety network to demonstrate these public safety applications. We have one transceiver site working in the Capitol Hill area. Please reference quickly the diagram to the left. These pictures taken yesterday show a real-time video teleconference between my team members located at the Capitol and MPD headquarters.

The solution leverages Flarion Technologies OFDM Flash Network and Motorola's newly developed greenhouse video dispatch application. The full 10-site network operating under an 18-month experimental license approved by the FCC is slated for completion in the summer of this year and will provide broadband wireless service throughout the District of Columbia. I would like to stress that the continuing cooperation from DHS and FCC is appreciated. We also enjoy our ongoing support of our corporate partners, Flarion Technologies, Televate, Motorola and SAIC.

In conclusion, please allow me to list some public safety challenges that this committee can help address. First, of course, is to provide the additional 10 megahertz of 700 spectrum for wide area broadband wireless public safety applications. Second, to accelerate, as we have mentioned earlier, the 700 megahertz spectrum clearing efforts. And last, of course, is to accelerate the resolution of the Nextel interference issue.

We look forward to the debate on the merits of our legislation, which we have included in our testimony, and we also welcome the opportunities to demonstrate this forward-thinking solution to the members of this subcommittee. I thank you for your support.

[The prepared statement of Robert Legrande follows:]

PREPARED STATEMENT OF ROBERT LEGRANDE, DEPUTY CHIEF TECHNOLOGY OFFICER, DISTRICT OF COLUMBIA GOVERNMENT

INTRODUCTION

Good afternoon, Mr. Chairman and members of the Subcommittee. My name is Robert LeGrande. I am a Deputy Chief Technology Officer in the Office the Chief Technology Officer (OCTO), the central information technology and telecommunications agency of the District of Columbia government. I am responsible for wireless communications infrastructure for the District government, and a representative of the Spectrum Coalition for Public Safety. I will describe for you how the District now has a state-of-the-art public safety voice network, complete with local, regional, and federal, interoperability and where we came from to get to this state. I will also describe the Spectrum Coalition for Public Safety's efforts to secure additional 700 MHz spectrum which will enable Public Safety to build and deploy Broadband Wireless Networks throughout the U.S. To reach this level of interoperability, we had to take several steps. First, we had to upgrade the coverage and capacity of our pre-existing non-interoperable local networks. Next, we had to unify these separate networks. Finally, we had to create interoperability between our intra-District public safety communications systems and other first responders in the region. We reached these goals by completing two major projects in September 2003 and March of this year. We have now embarked on the next step in fully loaded public safety communications capabilities: creating the high-speed broadband wireless data communications urgently needed by first responders throughout the nation. (Please see Attachment I, *Public Safety Wireless Voice and Data Communications*, for a graphic representation of these initiatives.) I will describe each of these efforts in grater detail, focusing particularly on the interoperability challenges we faced and the solutions we developed.

PUBLIC SAFETY VOICE COMMUNICATIONS IN THE DISTRICT OF COLUMBIA PRE-SEPTEMBER 2003

Before September 2003, the District's public safety radio communications infrastructure included two networks: a four-site Motorola SmartZone™ system operating at 800 MHz for Fire and Emergency Management Services (FEMS) and Emergency Management Agency (EMA) personnel, and a seven-site conventional analog system operating at 460 MHz for the Metropolitan Police Department (MPD). Both networks had major deficiencies. The FEMS network had insufficient in-building radio coverage in the core areas of the city—limitations compounded by the complex architecture of buildings in Washington, DC. (Despite these in-building coverage limitations, however, the network compared favorably with other major city networks in on-street coverage and quality.) There was no coverage in underground subway tunnels. The police network provided reasonable coverage throughout the city, but it was antiquated, failing, and in critical need of replacement. The network was over 30 years old, spare parts were no longer available from the original manufacturers, and some of them were no longer in business. Both networks suffered from capacity limitations. The FEMS-EMA 800 MHz network provided 16-radio channels, while the MPD UHF network had only 13 conventional channels and regularly experienced channel congestion intervals during the busiest hours. Our infrastructure had little to no interoperability within the District, due to the technical and operational disparity between the two networks, including frequency band and radio technology.

PUBLIC SAFETY RADIO COMMUNICATIONS UPGRADE

To solve these problems, a team of Motorola and District of Columbia engineers worked for six months to design an optimal unified communications network that would address the interlocking deficiencies of coverage, capacity, and interoperability in Washington, DC.

Coverage Analysis and Design

City management set an aggressive coverage goal of providing 100% communications within the District while minimizing the need for radio-to-radio communications (talk-around). We met this challenge in two projects, an above-ground project completed in September 2003, and an underground project completed in March 2004.

Our above-ground coverage analysis revealed that it was impractical to cover the interiors of all buildings using traditional radio sites. Instead, the analysis yielded a strategy to cover the majority (85%) of exterior and interior locations by expanding antenna sites from four to 10 and explore alternatives for covering the remaining areas. These alternatives were in—building distributed antenna systems and in-vehicle repeater systems. Our team quickly discovered that in-building systems were extremely expensive, created noise in the system that would degrade overall coverage, and could easily fail during fires or terrorist attacks. Vehicular repeater systems presented none of these problems, although they could not provide the same transparency as in-building systems, because they require first responders to change channels on their radios from the city-wide network to the vehicular repeater frequency. The city piloted a half-dozen vehicular repeater systems and found that single or multiple units could solve coverage problems in the densest of District buildings. Ultimately, the District implemented vehicular repeater systems in 63 fire suppression vehicles to ensure that a VRS would be available wherever needed to enhance in-building communications.

The subway tunnel system presented a more daunting challenge. The coverage gaps in tunnels were far too great to be addressed by VRS systems. However, sufficient resources existed underground to support a distributed antenna system. Therefore, the District, in partnership with The Metropolitan Area Transit Authority (WMATA) chose an underground distributed antenna system at 800 MHz and permitted the MPD to share WMATA's 490 MHz radio network that provides underground coverage. Key advantages of this system were the scope of coverage and transparency. Nearly 100% of all public underground areas were covered by the underground project completion date in March 2004,—there remains one lone corridor with fair voice quality will soon be upgraded to excellent voice quality.

Together, our above-ground and underground coverage solutions deliver nearly 100% coverage with only very limited need for radio-to-radio communication and provide District of Columbia first responders with citywide clear voice communication.

Interoperability and Capacity Analysis and Design

In addition to providing our first responders with the best possible radio coverage, we needed to deliver the best interoperability and capacity solution—the ability for District first responders to communicate within their agencies and among the maximum number of external agencies whenever necessary. Most of the District's surrounding counties use Motorola SmartZone™ technology[1] at 800 MHz. As discussed above, before the upgrade, the District had a seven site conventional analog system operating at 460 MHz for MPD and a four site Motorola SmartZone™ sys-

---

[1] For purposes of simplicity, we use SmartZone™ generically to describe both SmartNet™ and SmartZone™ systems.

tem operating at 800 MHz for FEMS and other District agencies. The District owned over 1,000 800 MHz radios compatible with the Motorola SmartZone™ system, nearly 2,000 portable 460 MHz radios with SmartZone™ capabilities and over 1,000 mobile 460 MHz radios capable of communicating on a SmartZone network. These same radios could be upgraded to support the public safety Project 25 radio standard, but not while maintaining important features and allowing dual-mode operations with SmartZone™ systems. Further, the surrounding municipalities operated mobile and portable radios that were programmed and configured to support SmartZone™ networks, but not Project 25 networks.

It is important to note that these radios operate in a single band. The 460 MHz radios operate in the 450—512 MHz range and the 800 MHz radios operate in the 806-824 MHz range. Therefore, a 460 MHz radio can not communicate directly on our neighboring county networks operating in the 806-824 MHz range. To alleviate this problem, the District aggressively sought to migrate MPD to 800 MHz. The team calculated a net requirement of 27-35 trunked voice channels at 800 MHz to satisfy aggregate demand for all District of Columbia public safety personnel. The District had 16 channels at 800 MHz and 13 channels at 460 MHz at the start of the analysis.

We considered several options for the migration:

- Implement additional 800 MHz frequencies,
- Use the public safety 700 MHz spectrum (24 MHz) and operate a 700/800 MHz network,
- Split the 16 existing 25 kHz channels to create up to 32 channels, and
- Create a dual-band 460/800 MHz network.

I'll review each option briefly.

Implement Additional 800 MHz Frequencies

To satisfy the aggregate demand, the District would need an additional 12 frequencies in the 800 Mhz band. Unfortunately, given the presence of our neighboring jurisdictions and Nextel in this band, we could not identify enough 800 MHz channels to meet our needs. We approached Nextel and engaged vendors to investigate short-spacing channels, both without success. Therefore, we had to discard this option.

Use the Public Safety 700 MHz spectrum (24 MHz) and Operate a 700/800 MHz Network

The additional channels in the 24 MHz of radio spectrum in the 700 MHz band presented some compelling opportunities. First, there were cost-effective multi-band radios on the market that could operate in both 700 and 800 MHz.[2] Second, there was considerable capacity in that band. Third, the technology used in the 700 MHz band, Project 25, was in the process of standardization, and therefore, presented an opportunity for expanded vendors and products. However, given the majority of users and systems operating SmartZone systems, our network needed to provide SmartZone service to agencies supporting District first responders within the city. Unfortunately, no integrated, dual-mode (P25 and SmartZone) network existed.

Moreover, the availability of the 700 MHz band was limited by the presence of TV broadcasters in our region. Therefore, we had to conclude that this option not feasible and halted efforts to build a Project 25-compatible network.

Split Existing 800 MHz Channels to Create up to 32 Total Channels

To implement this solution, a vendor would have to enable the use of adjacent channels at 12.5 kHz (instead of the existing 25 kHz) without interfering among the channels. Given the preponderance of SmartZone™ systems in the region, we first explored creating a SmartZone system that could utilize the half-spaced channels. Unfortunately, this option proved infeasible because the SmartZone system could not tune to those interspaced frequencies.

Create a Dual-Band 460/800 MHz Network

The dual band option would provide city-wide service from all sites at both bands and integrate them at a central hub. Analysis revealed that this option was not only feasible, but highly advantageous. It relied on existing frequencies licensed to the District of Columbia, and therefore presented limited risk of interference and licensing issues with the Federal Communications Commission (FCC). Motorola's existing SmartZone architecture could create a zone at 460 MHz and 800 MHz. This solution

---

[2] However, the entire police department would need new 800/700 radios, and FEMS might need new radios as well (their radios supported only 800 MHz). The result would be between 5,000 and 7,000 new 800/700 MHz radios costing $7-13 million more than the cost of upgrades to 460 MHz radios and new digital-capable 460 MHz radios.

could incorporate all of the existing 800 MHz radios and over half of the MPD radios. It also allowed adjacent agencies using 800 MHz radios to communicate with MPD radios at 460 MHz through the central hub. Further, by incorporating trunked radio technology, this solution delivered the maximum number of individual communications paths for simultaneous incidents. For example, this solution allows our first responders to communicate with Prince George's County Police while simultaneously maintaining a separate communications channel with United States Park Police but not consume resources when those channels were not needed. In addition, because WMATA uses a Motorola SmartZone network operating at 490 MHz, MPD could gain direct interoperability with WMATA and MPD will gain coverage within the tunnel system in July 2004. The dual-band option could also support a total of 27 trunked voice channels, providing adequate capacity for the first time.

The main disadvantage of this option was lack of interoperability for MPD officers operating outside the coverage area of our District of Columbia 460 MHz network. However, the disadvantage proves relatively insignificant. MPD officers travel outside our coverage area infrequently, as most mutual support situations (e.g., July 4th, Presidential Inaugurations, marches, and demonstrations) bring officers from neighboring municipalities into the District.

Upgrade Implementation

We implemented the coverage, interoperability, and capacity solutions I've just described on a fast track (April 2002-March 2004, less than two years from conception to completion) and at a relatively reasonable total cost of $42 million ($36 million in federal emergency preparedness funds, $2.5 million in grants, and $3.45 million in District funds). The results, as I've indicated, were overwhelmingly successful: we now have a full-coverage, 27 trunked voice channels, regionally interoperable system providing clear voice communication, encryption, and other digital features for all our first responders.

Of course, we faced numerous challenges along the way. We overcame these challenges through clear, unified direction and support from our Mayor, City Council, Deputy Mayor for Public Safety, Chiefs of Police and FEMS, Chief Technology Officer, and police and fire unions. In addition, we were fortunate in having strong, knowledgeable, and driven corporate partners, Motorola, Inc. and Televate, LLC.

RADIO INTEROPERABILITY WITHIN THE NATIONAL CAPITAL REGION/COUNCIL OF GOVERNMENTS

The National Capital Region (NCR) consists of in two states (Virginia and Maryland) and the District of Columbia. Voice radio interoperability for public safety entities in this region is essential. Equally essential for the District is interoperable communications with multiple critical federal agencies including the FBI, Secret Service, Bureau of Alcohol, Tobacco, and Firearms (ATF), Federal Emergency Management Agency (FEMA), the State Department, and others. There are also over 40 federal law enforcement agencies operating in the city, including Capital Police, Park Police, Mint Police and many others, with whom MPD dispatch and police officers must have direct communications. Finally, it is important that the District maintain communications within the WMATA subway tunnels and directly with police and airport authorities at the Reagan National Airport.

As illustrated in Attachment II *(Regional Public Safety Wireless Communications Interoperability)*, establishing voice radio interoperability with this wide array of agencies, many of which are operating multiple radio technologies in different regions of the radio spectrum, including VHF, UHF and 800 MHz, is a major technical, operational and administrative challenge. The interoperability cube in the attachment depicts the levels of interoperability planned by the region. The region continues to implement solutions to further enhance and simplify radio communications. More funding for technical and operational standards development and training, along with the installation of permanent, dedicated "interoperability" managers and technicians is required to ensure that these solutions remain readily available on demand in the community.

In order to simplify this complex radio communications effort, interoperability has been engineered into three levels.

Level One Interoperability: Spare incident radios (radio cache) operating on common interoperable channels, including mutual aid, are made available to local and national responders who do not have programmed UHF and 800 MHz trunked radios or conventional radios on regional mutual aid channels. The simplest, but not necessarily the most effective, means to achieve interoperability is to distribute on-location radios to incident commanders and responders. Existing radio caches and excess spare radio inventories within the District and NCR/COG are distributed as appropriate. In response to an identified shortage of spare radios in the NCR, the federal government provided a grant in FY 2004 to increase the availability of 800 MHz trunked radios. A 1,000 unit *COG Radio Cache* will be available beginning in mid-summer of 2004, just weeks away.

Level Two Interoperability: In order to achieve a higher level of interoperability within the NCR between separate public safety portable/mobile radios and telephone system exchanges, regional partners have implemented a "radio interface module" manufactured by JPS Communications, the ACU-1000. With assistance from the Department of Homeland Security (DHS) wireless division, SAFECOM, this technology has been successfully implemented in most of the jurisdictions and agencies (local, state and federal) in the region. The ACU-1000 device provides communication "patching" between agencies by integrating agency radios into an interface module. Radio patching allows dispatchers to manually facilitate radio communications between users of different technologies and frequencies. The District's ACU-1000 unit encompasses 21 distinct radios, supporting all local fire and police agencies and critical federal agencies.

Radio patching through the ACU-1000 or similar devices, while effective in enhancing interoperability, has various limitations and presents operational challenges. Agency radios must be integrated, maintained and programmed to reflect the latest radio user template. Since templates change almost annually for most public safety radio users, it is difficult to maintain up-to-date radios in the device. The technology also entails complicated set-up protocols, requires user training, and lacks standardized operational procedures. Because these devices are not daily equipment, end users can become "rusty" and function improperly. Because the networks are not integrated, this is the only means to connect multiple networks today.

Level Three Interoperability: The most effective route to interoperability for co-located work groups is to install directly compatible, same-technology systems and radios (trunked or conventional). Trunked networks, common in the NCR, must be programmed with common trunked system and radio IDs and interoperable talkgroups. Most of the fire department users in the region, except for Prince George's County in Maryland, have direct access to each other's 800 MHz trunked radio networks. When first responders in the region enter the city to assist the District's fire department, they can communicate on the District's radio network or vice versa. All users are operating on a common radio network using the same radio technology.

The new MPD radio network, while not at 800 MHz where surrounding county police reside, was designed to be fully compatible with local law enforcement radio networks through the use of a Motorola SmartZone radio network switch. The District is able to provide local law enforcement users access to the District 800 MHz trunked network, which supports direct communications with MPD radio users on their UHF network.

An alternative to direct radio network compatibility is to establish mutual aid channels for non-standard network users with call-in capability to a dispatch console. The District has implemented a conventional VHF channel that facilitates direct access for several federal agencies to the District's citywide MPD dispatcher. A federal user with this channel programmed into his/her radio can direct call the MPD dispatcher to request MPD support and/or communication with individual MPD officers. The District is now working with SAFECOM to enhance this mutual aid network, expand the number of usable channels to three, and extend coverage throughout the NCR. This approach will support regional interoperability between the District and federal user agencies and enhance interoperability among federal agencies and between federal users and surrounding NCR first responders. While not a perfect interoperability solution, the mutual-aid-channel design will provide near-term mobile communications between responder agencies.

Attachment III *(DC-Regional PS Voice Interoperability Status)* presents a tabular view of current and in progress interoperability within the NCR. This summary reflects the work of hundreds of public safety officials, first responders and technologists, who, with the support of Congress, dedicate their energy and lives to ensuring reliable and functional radio communications within the region and beyond. However, while our success to date is encouraging, we have more work to do to achieve simple, on demand regional and federal interoperability within the region. Public safety radios must be programmed directly to change talkgroups or frequencies. Therefore, while an interoperable network infrastructure exists, a considerable amount of work still remains to reprogram thousands of radios and train first responders how to use the new capabilities. Additionally, as discussed below, the Washington, DC NRC does not have interoperability with key Department of Defense agencies that is vital to higher-level emergency response.

INTEROPERABILITY BETWEEN THE DISTRICT AND DEPARTMENT OF DEFENSE AGENCIES

District officials and technologists have recently begun discussions with various Department of Defense (DoD) agencies to analyze the current state of interoperability between the parties. While the investigation is still in its infancy, hampered by lack of dedicated staff and capital resources, the results are clear: interoperability between NCR first responders and critical DoD agencies is insufficient and must be increased now to ensure that the affected agencies can meet near-term emergency communications requirements. The recommendations agreed upon between the DoD and NCR include implementing technical and operational solutions that are available today and expanding and institutionalizing the dialogue between the affected agencies to ensure that planned radio network changes and upgrades are regularly addressed and incorporated into the interoperability operations. It is important to note, however, that the District is already providing technical support to the Washington National Guard and has designed interoperability into a radio network enhancement that the Guard is now undertaking.

WIRELESS BROADBAND DATA NEEDS

The Challenge of High-speed Wireless Data Communication

The District's current wireless data communications capabilities rely on commercial cellular offerings at low speed (19.2 kbps). This speed provides extremely limited capabilities, largely restricted to text transmission. It also places public safety at risk from commercial networks that are not built to withstand long periods without power (e.g., hurricanes and winter storms) and lack enough redundancy to maintain connectivity between transceiver sites and central hubs. Additionally, the commercial technology upon which the District's public safety communications relies will be dismantled in 2005 forcing the District, and all such users nationally, to migrate to an alternative wireless transport technology.

Adequate response to emergencies ranging from multiple-alarm building fires to chemical, biological and other terrorist attacks requires immediate and rapid communications among multiple first-responders including fire, police, and emergency medical services. Currently, first-responders must rely on voice communications to receive time-sensitive information about an emergency incident. Information often comes too late or is lost altogether. Broadband wireless networks can dramatically improve public safety communications and operations nationally by providing full-motion, high-resolution video and other bandwidth-intensive monitoring tools to multiple first responders. These tools will allow sharing of time-critical information needed to respond more effectively to both routine and catastrophic events.

The demand on a wireless broadband network from one user can range from low-speed web browsing at 50-200 kilobits per second (kbps) to multiple real-time streaming video images transmitted at 1.2 megabits per second (Mbps). The District has demonstrated that its aggregate citywide demand on a network can exceed 50 Mbps and that usage can be concentrated in one area to require 10 Mbps per transmission site. Unfortunately, current public safety spectrum allocations at 700 MHz and 4.9 GHz for wireless data do not meet these needs, as data speeds do not meet individual and aggregate demand levels, or service is limited geographically and first responders must travel to hotspots to secure information—potentially losing critical life-saving time. Attachment IV *(Public Safety Spectrum Overview 1 and 2)* provides an analysis of the options available to public safety to satisfy high-speed wireless data needs.

At the root of the problem are radio propagation and channel bandwidth. The former results in signal degradation as the first responder travels farther from the transmission site (or when walls or other obstructions lie between the two endpoints). The latter results in decreased channel rates.

The propagation characteristics of radio frequency waves at 4.9 GHz and radio frequency waves at 700 MHz are so different that they result in extremely high deployment costs and operational costs for 4.9 GHz systems. In particular, as the transmitted frequency rises, the RF wave propagation transmission losses increase, thus reducing the coverage area of a base station. Therefore, assuming the deployment of the same technology, complete coverage of a city like Washington, DC would require significantly more sites at 4.9 GHz rather than at 700 MHz.

For instance, if we assume free space propagation conditions, all things besides the frequency considered being equal, the range of a 4.9 GHz base station would be seven times smaller than the range of a 700 MHz station.[3] Consequently, to provide citywide coverage would require almost 50 times the number of antenna sites at 4.9 GHz as at 700 MHz. The District of Columbia has estimated that about 420 sites would be needed to provide comprehensive coverage throughout the city at 4.9 GHz instead of the 10 required at 700 MHz, leading to significant deployment costs and prohibitive operational costs.

Actually, these comparisons are optimistic, as they are based on a free-space propagation assumption. In fact, the reality of the mobile propagation environment is worse, and actually worsens for higher frequencies. As described in a white paper published by TROPOS networks[4] natural or man-made obstacles generate propagation losses in addition to the free space propagation loss. In the referenced paper the authors compare 2.4 GHz to 4.9 GHz propagation characteristics. However, for the reasons explained above (propagation performance worsens as the frequency increases), the numbers in this paper would have to be considered lower bounds of propagation differences between 700 MHz and 4.9 GHz.

Those significant additional signal losses at the higher frequencies suggest that 50 to 100 *times* more sites would be needed for wireless coverage at 4.9 GHz to match coverage at 700 MHz. Thus, the 4.9 GHz spectrum is fundamentally limited in reach and requires numerous repeaters to reach even marginal distances. It is actually best suited to line-of-sight propagation, e.g. rooftop-to-rooftop communications, mesh-type networks where users can create a daisy chain for end-to-end communications, or short-distance communications around a fixed location (hot-spots).

Most public safety wireless data applications are expected to reach or support first responders wherever they are located in the District, whether driving car in a park or working in buildings. The 700 MHz band is the best-suited spectrum to support those applications.

Channel Bandwidth and Numbers of Channels

The maximum channel bandwidth in the existing 700 MHz allocation to public safety is 150 kHz. Technologies such as the standardized TIA-902 Scalable Adaptive Modulation have been tailored to this channel bandwidth and offer speeds up to 460 kbps. Unfortunately, this bandwidth does not support multiple video streams for an individual user. Furthermore, the 12 MHz[5] of radio spectrum set aside for wideband data must be shared among three states and over a dozen public safety agencies. Consequently, the District expects no more than three or four paired channels offering peak *citywide* throughput of 1.4 to 1.9 Mbps—far less than projected citywide demand and much less than aggregate demand for one transmission site.

Requirements for Broadband Wireless Data for First Responders

First responders need video, Geographical Information Systems (GIS), high-resolution still images, and other broadband data wherever their work takes them. On the highways, high-resolution images must be delivered as soon as possible. At the farthest points of any service area, first responders need to send and receive video for appropriate support. Further, first responders need broadband data delivered deep inside buildings on portable handheld devices, just as voice signals are now delivered by our new voice network. Table-1 below outlines the multiple requirements for broadband wireless data for first responders:

Table 1: Summary of Technical Requirements

General Requirements	
User Throughput	Designed to 80% load
Downlink (kbps)	1,500
Uplink (kbps)	500
Scalability	High, Minimal coordination burden when increasing capacity.
Mobility	Vehicular (>80 mph)
Coverage	Wide area (95% of Outdoor Area)
Connectivity	All IP addressable.

---

[3] The free space propagation at 1 km is 89.3 dB at 700 MHz, and 106.2 dB at 4.9 GHz. Those 17 dB propagation difference would result in a coverage radius ratio of 7 (coverage area ratio of 49), between the two frequency bands. Therefore obtaining the same services provided by the 10 sites covering the city at 700 MHz today would require more than 400 sites at 4.9 GHz.

[4] See http://www.troposnetworks.com/pdf/Spectrum_Whitepaper.pdf for further details.

[5] This represents the paired amount of spectrum for frequency duplexed operation. Of this 12 MHz, 5.4 MHz is reserved for future applications by the FCC. The total number of 150 kHz paired channels is 40.

Table 1: Summary of Technical Requirements—Continued

Cost	Comparable with existing cellular solutions.
Terminals	Supports standard device interfaces and offers low power consumption and small form factor options.
Large-Scale Incident Throughput Requirements	
Aggregate Demand (Entire District)	
Downlink (kbps)	56,100
Uplink (kbps)	20,080
Throughput Concentration	70% of major incident traffic in 20% of the city geography
Per Site Throughput (demand)	10 sites with the above throughput concentration
Downlink (kbps)	7,860
Uplink (kbps)	2,951
Per Site Throughput (with margin)	Designed to ~ 80% load
Downlink (kbps)	10,000
Uplink (kbps)	3,700
Net Capacity (Entire District)	
Downlink (kbps)	100,000
Uplink (kbps)	37,000

NATIONAL COALITION FOR PUBLIC SAFETY BROADBAND SPECTRUM

Recognizing that our wireless high-speed broadband data needs were the same as those of the rest of the nation, the District of Columbia founded the **Spectrum Coalition for Public Safety** (see Attachment V, *Spectrum Coalition Fact Sheet*). Thirty States, counties, cities, regions and public safety organizations quickly joined the Coalition. The public safety communications organizations documented their support in the attached letter (Attachment VI, *Public Safety Organization Support for New Broadband Spectrum Allocation*). The Coalition's objectives are to pursue legislation that would require the FCC to reserve 10 MHz of radio spectrum for wide area public safety broadband wireless uses; to enable competitive, affordable technologies that meet first-response requirements; and to facilitate nationwide network deployment. We have developed draft legislation (Attachment VII, *First Responders Enhancement Act (FREA)*) that calls for the spectrum allocation changes and have briefed more than 35 House and Senate member offices on our goals.

Design and Installation of Pilot Network

The urgent needs of first responders in the District of Columbia required more than pursuing legislation to facilitate network deployment. Our need is real and immediate. With the support of our public safety, technology, legislative, and executive leaders and our corporate partners—Motorola, Inc. and Flarion Technologies, Inc.—we obtained an experimental license from the FCC and are now installing the nation's first high-speed broadband wide-area wireless network for public safety. One additional partner, SAIC, is assisting us with application analysis. We have one live transceiver site and can transmit broadband radio signals throughout the Capitol Hill area. In late summer of 2004 we will complete installation of all 10 transceiver sites in the network and will provide broadband radio coverage throughout the District of Columbia. We will use the pilot to refine our system requirements for usability, scalability, reliability, and security. The applications planned for testing on the network include remote chemical and biological agent detection, video surveillance; helicopter video support, bomb squad video support, GIS applications, and EMS remote doctor support. This pilot network, with the full 10 MHZ allocation, will meet the requirements outlined in Table 1.

CONCLUSION

As the nation's capital, the District of Columbia faces unique and unusual public safety communications challenges. We have met the first level of these challenges by upgrading our public safety voice network to one of the best in the nation. We look forward to complementing that network with the nation's first citywide wireless broadband public safety network, and we hope that our leadership of the Spectrum Coalition will enable other jurisdictions to have the same public safety tools in the near future. We appreciate the support that the Coalition has received in both the Senate and House of Representatives and look forward to continuing our dialogue with the nation's leaders on the Coalition's critical objectives.

Mr. UPTON. Thank you.
Mr. Muleta? Welcome back.

STATEMENT OF JOHN B. MULETA

Mr. MULETA. Thank you. Good afternoon Chairman Upton, Ranking Member Markey and other members of the Subcommittee on Telecommunications and the Internet. I want to thank you for this opportunity to appear before you on behalf of the FCC to discuss our work in facilitating interoperability between the Nation's 40,000 public safety communications systems.

Under the leadership of Chairman Powell, the commission has intensified its efforts in this area and designated homeland security and public safety issues one of the commission's six core strategic objectives. As September 11th vividly demonstrated, the ability of public safety systems to communicate seamlessly at incident sites with minimal onsite coordination is critical to saving lives and property. The FCC is therefore committed to use all of its resources to promote and enhance the interoperability of the thousands of public safety systems that make up a critical part of our Nation's homeland security network.

Interoperability requires focus on more than spectrum, technology and equipment issues. It also requires focus on the organizational and personnel coordination and communication that is necessary to make it available in the time of our greatest need. For its part, the commission directs its efforts to, No. 1, providing additional spectrum for public safety systems; two, nurturing technological developments that enhance interoperability; and, three, providing its expertise and input for interagency efforts such as SAFECOM to improve our homeland security.

To date, specific FCC efforts have included designating blocks of spectrum between 100 and 900 megahertz for interoperability and emergency services; adopting regional planning as an alternative approach for spectrum licensing and management to drive coordination and communication, promoting and sharing of radio spectrum facilities, and adopting recommendations set by the Public Safety National Coordination Committee, exploring the potential of new technologies such as cognitive radios to enhance interoperability, and, finally, developing stronger day-to-day working relationship with SAFECOM and other critical organizations that help drive interoperability.

It is important to note that despite all of our efforts, there are limitations to what the FCC can do. The FCC is only one stakeholder in the process, and many of the challenges to interoperability exist because disparate governmental interests, local, State, and Federal, individually operate portions of our national public safety system. Each of these interests has different capabilities in terms of funding and technological sophistication, making it difficult to develop and deploy interoperability strategies uniformly throughout the country. Regardless of these problems, we at the FCC continue to advance policies that enable all of the stakeholders to do their best in maintaining a strong and viable national public safety system.

Turning to spectrum for public safety, the commission has currently designated throughout the country approximately 97 megahertz of spectrum from 10 different bands for public safety use. Public safety entities also actively use spectrum-based services in other spectrum bands.

For example, under the ultra-wideband rules, the ground penetrating radars and imaging systems enable public safety users to detect the location or movement of people behind or within walls or other structures, an important and potentially lifesaving tool. Moreover, the available priority access services on some commercial wireless networks gives certain emergency personnel greater ability to access commercial, cellular and personal communication services in times of crises.

Looking at more recent public safety spectrum allocations, in the last few years, the commission has made two allocations that illustrate the importance placed on assuring that public safety entities have the sufficient spectrum to carry out their critical missions. First, consistent with the Balanced Budget Act of 1997, the FCC identified and allocated 24 megahertz of spectrum in the 700 megahertz band for public safety used, as has been noted by many folks today.

As part of this proceeding, the FCC dedicated 2.6 megahertz of this spectrum for interoperability purposes. Given the central role that states play managing emergency communications, the FCC also concluded that the states are best suited for administrative interoperability spectrum and that State level administration will promote safety of lives, property through seamless coordinated communications on interoperability spectrum.

The FCC also designated 50 megahertz of spectrum at 4.9 gigahertz for public safety users in the response to requests from the public safety community for additional spectrum for broadband data communications. The 4.9 gigahertz band will also foster interoperability in two ways: One, by providing a regulatory framework where traditional public safety entities can license it on a shared basis and where they can also pursue strategic partnership with other non-public safety actors as needed for the completion of their mission.

In addition to using its resources to identify additional spectrum, the FCC has also provided for, No. 1, innovative licensing methods; two, creating planning methods that encourage better coordination and communication; and, No. 3, promoted new technologies. Foremost in this area, the commission adopted the regional planning approach to spectrum management as an alternative to the traditional long-held belief in first-in-the-door approach to spectrum licensing and management in the public safety context.

In order to promote interoperability, the commission also permits 2 types of spectrum sharing. First, the FCC's rules specifically provide for shared use of radio stations where licensees may share facilities on a non-profit, cost-shared basis with other public safety organizations and end users. In July of 2000, the commission expanded this sharing provision. This rule also allows Federal Government entities to share these facilities as end users.

A second type of sharing is unique to the 700 megahertz public safety spectrum. In this band, State and local public safety licensees may construct and operate joint facilities with the Federal Government. The commission took this action to encourage partnership of FCC-licensed State or local government entities with Federal entities in order to promote interoperability and more efficient use of the spectrum.

To promote the new technologies, the FCC chartered the Public Safety National Coordination Committee, NCC, which operated as a Federal advisory committee between 1999 and 2003. The NCC recommended technical and operational standards to assure interoperability in the 700 megahertz public safety band. The NCC worked with the Telecommunications Industry Association, a credited standard developer, to develop interoperability technical standards that are open and non-proprietary.

Moving on to the coordination issue, the FCC recognizes interagency coordination as an essential factor in developing effective interoperability. To that end, my staff and other staff of the FCC routinely confers with critical organizations, including APCO, the National Public Safety Telecommunications Council, the International Association of Fire Chiefs and International Association of Chiefs of Police.

Moreover, my staff has been working closely with the Department of Homeland Security's SAFECOM. The FCC and SAFECOM share the common goal of improving public safety communications interoperability. We are continuing to work on our collaborative efforts to develop a strong working relationship, both formally and informally.

For example, the FCC is an active member of SAFECOM's Advisory Group. In addition, FCC staff meets routinely with staff from SAFECOM, including on several occasions where information was exchanged and we received briefings. Most recently, we did this on a March 11 presentation to SAFECOM's Executive Committee on matters pending before the commission. The FCC has also attended and participated in several events hosted by SAFECOM, including its 2003 Summit on Interoperable Communications for Public Safety and the 2004 Public Safety Communications Interoperability Conference.

Moreover, on a personal level, Dr. Boyd and I have established direct lines of communication between us to promote and ensure effective coordination regarding homeland security and public safety communications initiatives.

I would like to thank you for the opportunity to testify in front of you on this important issue affecting our homeland security, and I will gladly answer any questions you might have. Thank you.

[The prepared statement of John B. Muleta follows:]

PREPARED STATEMENT OF JOHN B. MULETA, CHIEF, WIRELESS TELECOMMUNICATIONS BUREAU, FEDERAL COMMUNICATIONS COMMISSION

INTRODUCTION

Good afternoon Chairman Upton, Ranking Member Markey and other Members of the Subcommittee on Telecommunications and the Internet. Thank you for this opportunity to appear before you on behalf of the Federal Communications Commission to discuss our work in facilitating interoperability between the nation's public safety communications systems.

Currently, there are more than 40,000 spectrum licenses designated for public safety systems under the Communications Act. The Commission has the unique role of providing spectrum for state and local governments to use as part of these systems. As a result, the Commission has had a long-standing commitment to the protection and enhancement of public safety communications systems. Under the leadership of Chairman Michael K. Powell, the Commission has intensified its efforts in this area and designated homeland security and public safety issues one of the Commission's six core strategic objectives. As September 11, 2001 demonstrated, the ability of public safety systems to communicate seamlessly at incident sites with minimal on-site coordination is critical to saving lives and property. The FCC is therefore committed to use all of its resources to promote and enhance the interoperability of the thousands of public safety systems that make up a critical part of our nation's homeland security network.

The Commission's experience indicates that a holistic approach is the best method for fostering interoperability. Achieving interoperability requires an emphasis on more than spectrum, technology and equipment issues—it also requires a focus on the organizational and personnel coordination and communication necessary to make interoperability available in times of greatest need. For its part, the Commission directs its efforts toward providing additional spectrum for public safety systems, nurturing technological developments enhancing interoperability and providing its expertise and input for interagency efforts such as SAFECOM.

There are limitations, however, to what the FCC can do. The Commission is only one stakeholder in the process and many of the challenges facing interoperability are a result of the disparate governmental interests—local, state, and federal—that individually operate portions of our national public safety system. Each of these interests has different capabilities in terms of funding and technological sophistication, making it difficult to develop and deploy interoperability strategies uniformly throughout the country. Regardless of these problems, we at the FCC continue to advance policies that enable all of the stakeholders to do their best in maintaining a strong and viable national public safety system.

COMMISSION RESOURCES

The FCC works in an integrated and flexible fashion to assign spectrum for public safety purposes. The Wireless Telecommunications Bureau (WTB) and the Office of Engineering and Technology (OET) share significant responsibility for intra-agency projects related to interoperability technology and policy development. The Commission also maintains a Homeland Security Policy Council (HSPC) and created the Office of Homeland Security within the Enforcement Bureau to facilitate intergovernmental communications on homeland security issues.

Wireless Telecommunications Bureau

WTB underwent a reorganization this past year that created the Public Safety and Critical Infrastructure Division (PS&CID). PS&CID now has a clear focus—its job is to administer the licensing rules for public safety radio networks and the related radio networks of critical infrastructure industries such as the nation's utilities. The division also has the responsibility of promulgating rules that require wireless carriers to deploy E911 systems throughout the country for the benefit and use of over 160 million cell phone subscribers—another critical element of the nation's homeland security system. The division's routine day-to-day contact with public safety licensees, their vendors and other stakeholders allows it to closely monitor industry trends and needs. In 2003, WTB processed more than 529,000 public safety and other private and mobile applications, including applications for new licenses, license modifications and renewals, waivers, and requests for special temporary authority.

Office of Engineering and Technology

In addition to its responsibility for spectrum allocations, OET routinely assesses vulnerabilities in communications networks and equipment and makes recommendations for facilitating improvements to network security, reliability and integrity. OET also evaluates new technologies and makes recommendations to the Commission for rule changes which would enable their use to improve the communications capability of the nation's public safety community. OET is the agency's principal point of contact with the National Telecommunications and Information Administration (NTIA) and in this role works with NTIA on spectrum issues that affect both non-Federal and Federal government spectrum users, including state, local and federal first responders.

Homeland Security Policy Council and Office of Homeland Security

The FCC's Homeland Security Policy Council (HSPC), created in November, 2001 and composed of senior managers of the Agency's policy bureaus and offices, and the Office of Homeland Security (OHS) assist the Commission in implementing the Homeland Security Action Plan. Among the directives of the Action Plan is to ensure that public safety, public health, and other emergency and defense personnel have effective communications services available to them as needed.

Equally as important, HSPC and OHS ensure coordination with other federal, state, and local entities that are involved with Homeland Security. For example, as a partner with the Department of Homeland Security, the FCC has promoted registration of states and localities in the Telecommunications Service Priority and the Wireless Priority Access Service programs. These programs provide wireline and wireless telephone dial tone to public safety entities on a priority basis during and following a disaster. HSPC members are also working with disabilities rights organizations to identify and resolve communications issues that have an impact on that community during national emergencies.

In addition, HSPC and OHS work closely to support the Network Reliability and Interoperability Council (NRIC VII) and Media Security and Reliability Council (MSRC), two of the FCC's federal advisory committees. Through NRIC VII, communications industry leaders provide recommendations and best practices to the FCC focused on assuring optimal reliability and interoperability of wireless, wireline, satellite, paging, Internet and cable public communications networks and the rapid restoration of such services following a major disruption. MSRC does much the same with the goal of achieving optimal reliability, robustness and security of broadcast and multi-channel video programming distribution facilities. Public safety representatives are part of this effort since, during emergencies, TV and radio are sources of information for citizens.

SPECTRUM FOR PUBLIC SAFETY

The Commission currently has designated throughout the country approximately 97 MHz of spectrum from ten different bands for public safety use. Public safety entities also actively use spectrum-based services in other spectrum bands. For example, under the ultra-wideband rules, ground penetrating radars and imaging systems enable public safety users to detect the location or movement of people behind or within walls or other structures, an important and potentially lifesaving tool. In addition, various frequencies are available from 2 to 25 MHz for emergency communications.

The Commission also grants licenses to public safety entities for non-public safety spectrum to promote effective and efficient public safety communications. Such actions have allowed public safety entities to implement state-of-the-art communications systems and/or increase interoperability. Also, many public safety entities use commercial wireless communications to supplement their other non-emergency communications. Finally, the availability of Priority Access Service (PAS) on some commercial wireless networks gives certain emergency personnel greater ability to access commercial cellular and Personal Communications Service (PCS) systems in times of crisis.

Spectrum Dedicated for Public Safety Interoperability

The Commission has designated certain channels in the public safety bands for public safety interoperability. A public safety entity may use these designated frequencies only if it uses equipment that permits inter-system interoperability. The frequencies that have these so-called "use designations" include 2.6 MHz of the 700 MHz band, 5 channels in the 800 MHz band, 5 channels in the 150 MHz band (VHF Band), and 4 channels in the 450 MHz band (UHF Band).

Starting on January 1, 2005, the Commission will require newly certified public safety mobile radio units to have the capacity to transmit and receive on the nationwide public safety interoperability calling channel in the UHF and VHF bands in which it is operating. Also, in the case of certain inland coastal areas, known as VHF Public Coast areas (VPCs), the Commission has designated several additional channels in the VHF band to be used exclusively for interoperable communications.

Recent Public Safety Spectrum Allocations

In the last few years, the Commission has made two allocations that illustrate the importance placed on ensuring that public safety entities have sufficient spectrum to carry out their critical missions. First, consistent with the Balanced Budget Act of 1997, the Commission identified and allocated 24 MHz of spectrum in the 700 MHz band for public safety use. Second, the Commission made available for public safety use 50 MHz of spectrum at 4.9 GHz.

To better facilitate use of the 700 MHz public safety spectrum, the Commission adopted special rules and policies. It crafted provisions both to address the continuing interoperability issues among various public safety systems and to provide flexibility to accommodate a wide variety of innovative uses. In particular, the Commission dedicated 2.6 MHz of this spectrum for interoperability purposes. Given the central role that states provide in managing emergency communications, the Commission concluded that states are well-suited for administering the interoperability spectrum and that state-level administration would promote safety of life and property through seamless, coordinated communications on the interoperability spectrum.

The FCC's rules provide that the states may manage interoperability channels in two ways: (1) they may establish a State Interoperability Executive Committee (SIEC) or its equivalent; or (2) they may designate their Commission established Regional Planning Committees (RPCs). Thirty-eight states and the District of Columbia elected to administer their interoperability spectrum. For the fourteen that did not, the RPCs have been delegated the responsibility to administer this spectrum.

From the beginning, the Commission has recognized that the utility of this spectrum for public safety depended on taking actions, consistent with the current statutory scheme, to minimize, and ultimately clear, the broadcast use of this spectrum. For instance, during the digital television ("DTV") planning, the Commission minimized the use of channels 60-69. As a result, the new 700 MHz public safety spectrum on TV channels 63-64 and 68-69 is available now in many areas of the country. Because of the significance of this spectrum for public safety, especially first responders and interoperability, the Commission is actively considering ways to bring the digital transition to its conclusion. Indeed, under the direction of Chairman Powell, the Media Bureau has developed a bold framework that would provide a soft landing and a clear conclusion for the DTV transition so that, in part, we can provide public safety with this additional spectrum.

The Commission's second allocation, 50 MHz of spectrum at 4.9 GHz (4940-4990 MHz), promises to permit the use of new advanced wireless technologies by public safety users. This spectrum is part of a transfer of Federal Government spectrum to private sector use. The Commission initially proposed to allocate the 4.9 GHz band for fixed and non-aeronautical mobile services and to auction it to commercial users, with no designation of the spectrum for public safety use. In response to requests from the public safety community for additional spectrum for broadband data communication, the Commission designated the 4.9 GHz band for public safety use in February 2002 and adopted service rules in April 2003.

The Commission intended the 4.9 GHz band to accommodate a variety of new broadband applications such as high-speed digital technologies, broadband mobile operations, fixed "hotspot" use, wireless local area networks, and temporary fixed links. The 4.9 GHz band rules also foster interoperability, by providing a regulatory framework in which traditional public safety entities can pursue strategic partnerships with others necessary for the completion of their mission.

Licenses for this spectrum will be granted to public safety entities based on a "jurisdictional" geographical licensing approach. Accordingly, the 4.9 GHz spectrum will be licensed for shared use. Under this approach, the Commission will authorize 4.9 GHz licensees to operate throughout those geographic areas over which they have jurisdiction and will require them to cooperate with all other 4.9 GHz licensees in use of the spectrum. In order to increase spectrum use and foster interoperability, the Commission will permit licensees to enter into sharing agreements or strategic partnerships with both traditional public safety entities, including Federal Government agencies, and non-public safety entities, such as utilities and commercial entities.

PROMOTION OF PUBLIC SAFETY INTEROPERABILITY

There are a range of mechanisms that specifically promote interoperability. As discussed above, the Commission has used its resources to identify additional spectrum. The Commission also has provided for innovative licensing methods, created planning methods that encourage better coordination, and promoted new technologies.

Regional Planning

The Commission adopted the regional planning approach to spectrum management as an alternative to the traditional first-in-the-door approach to spectrum licensing and management in the public safety context. Regional planning allows for maximum flexibility of the RPCs to meet state and local needs and encourage innovative use of the spectrum to accommodate new and as yet unanticipated developments in technology and equipment. The Commission has utilized this approach for public safety spectrum in the 700 and 800 MHz bands.

Sharing of Radio (Spectrum) Facilities

In order to promote interoperability, the Commission has rules for two types of spectrum sharing. First, the FCC's rules specifically provide for shared use of radio stations where licensees may share their facilities on a nonprofit, cost shared basis with other public safety organizations as end users. In July 2000, the Commission expanded this sharing provision. This rule also allows Federal government entities to share these facilities as end users. A second type of sharing is unique to the 700 MHz public safety spectrum. In this spectrum band, state and local public safety licensees may construct and operate joint facilities with the Federal government. The Commission took this action to encourage partnering of FCC-licensed state or local government entities with Federal entities to promote interoperability and spectrum efficiency.

Public Safety National Coordination Committee

The Public Safety National Coordination Committee (NCC) operated as a federal advisory committee from 1999 to 2003 and recommended technical and operational standards to assure interoperability in the 700 MHz public safety band. The over 300 members employed a consensus-based decision-making process to meet its charge. The NCC was guided by an eleven-member Steering Committee and used three subcommittees, each of them having several working groups to develop its recommendations, many of them highly technical. It submitted its final recommendations in July 2003.

The NCC developed recommendations on a technical standard for the narrowband voice and data channels to ensure that police, firefighters, EMS and other public safety officials using 700 MHz radios can communicate with one another instantly on common voice and data channels. The same channels are designated for interoperability use everywhere in the United States. The Commission adopted the narrowband voice standard and also a narrowband data standard in January 2001 as the NCC recommended.

The NCC also developed a recommendation for a wideband data standard and forwarded it to the Commission in July, 2003. This standard would give public safety agencies a common "pipeline," on 700 MHz wideband data interoperability channels, with which to implement such applications as sending mug shots and fingerprints to police vehicles, medical telemetry from EMS units to hospitals, blueprints of burning buildings to firefighters and video coverage of incidents to the incident commander. The NCC worked with the Telecommunications Industries Association—an accredited standards developer—to develop interoperability technical standards that are open and non-proprietary. The Commission will consider the remaining NCC recommendations, including the wideband data standard, in a future rulemaking.

Intelligent Transportation Systems Radio Service

In December 2003, the Commission adopted service and licensing rules for the Dedicated Short Range Communications (DSRC) Service in the Intelligent Transportation Systems (ITS) Radio Service in the 5.850-5.925 GHz band. It is envisioned that DSRC would provide the critical communications link for ITS, which is key to reducing highway fatalities, a high priority for the Department of Transportation. The effective and expeditious implementation of DSRC not only benefits American consumers by providing solutions to today's transportation challenges and allowing life-saving communications. It also provides public safety entities with another communications tool that can assist them in fulfilling their missions. To ensure interoperability and robust safety and public safety communications among DSRC devices nationwide, the Commission adopted rules requiring that the ASTM-DSRC standard be used. The Commission also adopted licensing and technical rules aimed at creating a framework that ensures priority for public safety communications, thereby allowing both public safety and non-public safety use of the 5.9 GHz band. Further, the Commission adopted a jurisdictional licensing approach similar to that used for the 4.9 GHz band.

Cognitive Radios Proceedings

The Commission is actively exploring the potential of new technologies to enhance interoperability and encourage network efficiency of public safety systems. One example of such new technologies is cognitive radios, which have the capability to change their power and/or frequency, sense their environment, know their location, and optimize their communication path. This technology holds tremendous promise for public safety interoperability by making it possible for radios from different public safety systems to operate seamlessly at an incident site without prior coordination. The Commission has initiated a Cognitive Radio Technologies proceeding to examine the enhanced interoperability potential that these even more flexible technologies may offer.

COORDINATION

The FCC recognizes that interagency coordination is an essential factor in developing effective interoperability. To that end, Commission staff routinely confers with the Department of Homeland Security's SAFECOM. The FCC and SAFECOM share the common goal of improving public safety communications interoperability. We are continuing our collaborative efforts to develop a strong working relationship, both formally and informally. For example, the FCC is an active member of SAFECOM's Advisory Group. In addition, FCC staff has met with staff from SAFECOM on several occasions for information exchanges and briefings, including, most recently, a March 11, 2004 presentation to SAFECOM's Executive Committee on matters pending before the Commission.

FCC staff also has attended and/or participated in several events hosted by SAFECOM, including its 2003 Summit on Interoperable Communications for Public Safety and 2004 Public Safety Communications Interoperability Conference. Moreover, DHS Deputy Director David Boyd and I continue to work together to further promote and ensure effective coordination regarding homeland security and public safety communications initiatives. We agree that it is critical that the FCC and SAFECOM continue to work cooperatively to achieve our common interests of promoting homeland security and interoperability.

CONCLUSION

The FCC is dedicated to marshalling all of its resources and expertise in order to ensure that adequate spectrum and technology is available for providing interoperability among the nation's public safety systems. The Commission continues to work with a wide range of stakeholders to foster and promote new policies, rules, regulations and technologies related to public safety interoperability. Although some of the challenges involved in bringing interoperability to public safety systems are outside the scope of the FCC's authority, the Commission continues to take a leadership role in trying to resolve these challenges.

Thank you for the opportunity to testify on this important issue affecting our homeland security.

Mr. UPTON. Well, thank you much, all four of you. At this point, we will go to questions from members of the panel.

Mr. Grube, I want to say I appreciated very much, your testimony, particularly as you referenced that there was no hard date on the transition, that you are in limbo. I want to say I know that I speak for the chairman, who will be asking questions soon, that we do want a hard date, and we do want people to know when that date will be, and we intend next month to have a hearing, yet another hearing on the transition, specifically on the Berlin model and what we can learn from their experience.

Mr. LeGrande, those of us that share DC as a second home, for those of us that commute from our states, Mr. Stupak and me from Michigan, we appreciate the work that you have done to upgrade our city's resources here, and I have a couple of questions. You indicated in your testimony that the District firefighters had pretty good interoperability, being able to communicate both above and below ground with the subway system. Do the police have that same capability? What about EMS? And what about their ability to communicate with each other in those same scenarios?

Mr. LeGrande. Okay. First, the fire department does have seamless communications, meaning they don't have to change their radio channels or anything like that when they go from above or below ground——

Mr. UPTON. I am actually a member of the Firefighter Caucus, and actually there was 1 day, not too many years ago, that we actually rode with the department. They didn't have that capability then.

Mr. LeGrande. March of this year, that is when they got it. And the police department's upgrade will be completed in July of this year.

Mr. UPTON. Will they be able to communicate with each other then as well?

Mr. LeGrande. Yes. We have intra-District—we call it intra-District interoperability where our police, fire and EMS do have the capability above ground to interoperate now. When the police come on in the subway system, they too will be able to interoperate as they do now above ground.

Mr. Upton. Did you experience cultural challenges, disputes between the two departments?

Mr. LeGrande. Yes.

Mr. Upton. You have not sworn under oath, but we want your honest answer.

Mr. LeGrande. I do have to go home after I leave here.

Mr. Upton. They are outside the door waiting for you.

Mr. LeGrande. Yes, absolutely. But, you know, first, when I went through the process, honestly, I understood. I paused to understand that over 30 years we have developed these systems and just understanding based on a finite set of requirements and a threat that is usually jurisdictional. So I kind of understood that there was a reluctance on some parts to do that.

But I think what I have found in our first responders, not only here but also in the various first responders that we have met through the Public Safety Spectrum Coalition, that there is a commitment on their part, and I think that sometimes the difficulty in culture is somewhat—I will just say somewhat exaggerated. I know there are cases where it isn't, but some cases they are very willing to work together to help this communications problem.

Mr. Upton. Dr. Boyd, you referred to their cultural challenges between different departments. What do you see that we have to do to overcome some of those challenges?

And, Mr. Muleta, I would like you to respond to that too.

Mr. Boyd. I think your insight that the cultural issue is a critical piece of interoperability is on the mark. Our experience has been that there is an increasing interest on the part of all the disciplines in actually communicating with each other and jurisdictions in communicating with each other. When you get to the details, it is sometimes fairly difficult because it begins to threaten existing structures.

We are finding increasing levels of cooperation, however, interdisciplinary as well interjurisdictionally. And one of the things that we have discovered as crucial in creating interoperability is a governance structure that works from the lowest level up. Our experience has been that Federal interoperability efforts tend to fail because we try to drive them too often the top instead from working from the bottom. The same thing happens at the State level. And if you can build a really good model that starts with the most local level and work up, then you can begin to really resolve interoperability.

We had a project recently with the State of Virginia, we will be producing a report shortly, where we worked with them to experiment with exactly that model in the development of a statewide plan, which we think is working out really well. And we started that with the most rural, smallest jurisdictions in the State and then worked our way around. That, we think, is the key to fixing the cultural issues.

Mr. Upton. Mr. Muleta?

Mr. MULETA. I do think cultural issues exist, but at the FCC what we have tried to do is a couple of things. One is we are spending a great deal of time with the various public safety communities, both at the Federal and the State level, to sort of understand the requirements and through the regional planning process sort of define a common set of issues and then work around those. So I think that has been incredibly helpful.

Inside the FCC we have also make great strides in making sure that there are no walls between various parts of the FCC. Chairman Powell has created the Office of Homeland Security and there is a Homeland Policy Council as well as within my organization I recently reorganized to put in all of the elements of public safety issues, including E911, which really plays an important role in sort of threat identification and management, to be part of our overall look in public safety.

So between better coordination, a more holistic understanding and planning of issues, we are trying to address these issues, and I think, unfortunately, the events of the last few years where the threat, as you mentioned, have been much greater, have helped make all of us realize that we have to work together.

Mr. UPTON. Mr. Stupak?

Mr. STUPAK. Thank you, Mr. Chairman, and thanks for holding this hearing. I ask that my full statement be made part of the record.

Mr. UPTON. Without objection, all members' statements will be made part of the record.

Mr. STUPAK. Thank you, Mr. Chairman. We had a hearing earlier this month where we discussed basically the same challenges, and I would be interested in knowing what has happened since then. From what I can see, not much has happened. We still have the same challenges, there is no funding, we still have the spectrum interference, as we are hearing about, we still don't have any real coordination plan in meeting this goal.

It has been almost 3 years since 9/11 and I really don't see a lot happening and I am really disappointed we didn't have at least some first responders here today to tell us what they are hearing on the street, because while we have all these offices and new policies, even in a seamless radio connection like we have here at DC, if the police officer or the fire department individual who's out of his car, out of the station cannot respond back and talk to each other, it doesn't do us a whole heck of a lot of good.

So let me ask Mr. LeGrande, in your seamless radio system here, can a police officer outside his car talk back to stations, thing like that, on his hand-held?

Mr. LEGRANDE. Yes.

Mr. STUPAK. All right. Can he talk to a fire department official?

Mr. LEGRANDE. Yes.

Mr. STUPAK. Okay. Can he talk to the Capitol Police?

Mr. LEGRANDE. Yes.

Mr. STUPAK. How about the Park Police?

Mr. LEGRANDE. In my testimony, there is a detailed status in the attachment 3——

Mr. STUPAK. Sure.

Mr. LeGrande. [continuing] and there are varying statuses of where we are. Now, as far as the technical aspects of it, we put in a system that will allow it. Currently, what we are doing is working through the standard operations procedures. So the technology exists. We are working through the process to——

Mr. Stupak. How many agencies do we have just in DC here alone? Don't we have like about 25 to 30 different agencies?

Mr. LeGrande. Twenty-four in the region.

Mr. Stupak. Twenty-four. Can we all talk to each other?

Mr. LeGrande. From the DC perspective and from the DC police and fire, we have the capability to talk to each one of those. We are working on the finalizing the process. Before you can actually go out and implement that capability, you really have to go through a process definition and then a very detailed training. So we are in the process of doing that. We are well on our way.

Mr. Stupak. So a firefighter out of his wagon there, or whatever you want to call it, he is in the building, he can't talk to other members from other agencies yet. That is still not there.

Mr. LeGrande. The capability exists, yes, for him to talk to other agencies. And if he needs to right now, there is a process of even patching him through right now. So I guess what I am giving you a status of is that there is an ability for us to go—for him to speak——

Mr. Stupak. I don't want to dispute it as an ability, but can they actually do it? Are the actually doing it? I guess that is what I am asking. It is almost 3 years now, and we saw $100 million in the budget in fiscal year 2003 for a $6 billion to $8 billion problem. So we put $100 million in the budget and that is been it.

So I guess what I am trying to get at here today, I have heard a lot about abilities and robust planning and all this, but I mean this has been going on for a long time.

I have been associated with law enforcement for 30 years. This has been going on for 30 years, and we still don't have it. I am not blaming you guys. I am just maybe voicing a little frustration, but I just really think that we really have to get at this and allow that officer on the street or that emergency medical person to talk to whoever they need to talk to and not have to worry about having it patched back through dispatch and dispatch then patch it back to somebody else.

Mr. LeGrande. Okay. Well——

Mr. Stupak. And that is what I am trying to get at.

Mr. LeGrande. All right. Let me try to specifically answer your question. With regards to the fire department, they have currently the ability to speak to Washington Airport Authority, Fairfax County, Fairfax County Police Department, Alexandria Fire Department, Alexandria Police Department, the Arlington Police Department and the Arlington Fire Department right now.

Mr. Stupak. But can they talk to each other? I guess that is what I am really asking.

Mr. LeGrande. Absolutely.

Mr. Stupak. Okay.

Mr. LeGrande. I am sorry. Maybe I misunderstood your question.

Mr. STUPAK. So the command officer on the street can talk to the guy up in the building and tell him what is going on.

Mr. LEGRANDE. Yes, sir.

Mr. STUPAK. So we don' t have the thing that happened at World Trade Center where those people don't know what is going on.

Mr. LEGRANDE. There are two questions you are asking——

Mr. STUPAK. Sure.

Mr. LEGRANDE. [continuing] and I will just separate the two. There are 2 problems on 911: Interoperability and in-building coverage.

Mr. STUPAK. Correct.

Mr. LEGRANDE. Okay. If you are in the District of Columbia, that couldn't be more underscored with the marble buildings that we created here and we had to build that in our design. We put in a new 10-site system which increased the coverage and capacity within the District. We also added 63 vehicle repeater systems, such that if there is a major incident, we can go and deploy these vehicles which will get around building penetration radio signal. So, yes, they can absolutely speak.

Mr. STUPAK. So in order to talk to them, they have to have that repeater vehicle there.

Mr. LEGRANDE. Only if it is a very thick building requiring that, and we know where those buildings are. And, by the way, those units are some of the first units that are deployed.

Mr. STUPAK. Do you know how much money you have spent on this system to try to get it to where it is at today?

Mr. LEGRANDE. Approximately $42 million.

Mr. STUPAK. $42 million. And did that come from the Federal Government?

Mr. LEGRANDE. Yes, sir.

Mr. STUPAK. So of the $100 million we have spent, $42 million went to DC?

Mr. LEGRANDE. I believe that is the case. I can't answer that question.

Mr. STUPAK. Just sort of magnifies the need across the Nation.

Mr. LEGRANDE. Yes, it does, sir.

Mr. UPTON. Mr. Barton?

Chairman BARTON. Thank you, Mr. Chairman, and thank you for holding this hearing. I want to thank the panel for being here.

My first question is fairly elementary, but I want to make—all these interoperability channels that we are talking about, are these channels that only the first responders and law enforcement officials have access to or can anybody with a police scanner or monitor listen in on these channels?

Mr. LEGRANDE. Yes. These are channels that can be monitored.

Chairman BARTON. So anybody that—a terrorist, if they took the time to go to Wal-Mart, could get a scanner and hearing everything that was going on.

Mr. LEGRANDE. We, within the design of our system, at least within the District of Columbia, included a significant amount of encryption, which would prevent sensitive communications from being monitored.

Chairman BARTON. What does that mean in plain language?

Mr. LeGRANDE. That we have the ability to make the signal between—an encrypted between one person and the other where they couldn't be scanned.

Chairman BARTON. They would just hear static.

Mr. LeGRANDE. Yes, or nothing at all.

Chairman BARTON. How often is that actually done?

Mr. LeGRANDE. Well, right now there hasn't been any, I believe—I really can't speak to that how often it has been done, but the capability exists and is planned to be used in an incident where we have to communicate sensitive information.

Chairman BARTON. Okay. That leads to my next question, and I don't know that this would be possible, but given the fact that most law enforcement communication equipment can be scanned, would it be possible to use some sort of a special cell phone or even a regular cell phone that had special priority, so that in the case of an emergency you could code a certain code into the cell phone and those calls would go through first and get priority? Because you cannot or it is very difficult to monitor a cell phone call. Is there any possibility to use some sort of a system like that?

Mr. MULETA. If I can address that question, I think, first of all, the general question—maybe Mr. Grube can also address this—is when you move to digital communications, it is much easier to encrypt and therefore protect communications even if it is on a public safety radio system. So part of the transition that we have all been talking about is moving to a uniform standard that has interoperability and enables visual communications.

Part of the transition process is to be able to upgrade, you know, uniformly throughout the country all the systems so that, you know, they receive the benefits of encryption and various things like that. I think the second question you asked is is there a way of providing what is known as priority access, that is, a program that we have been working with folks at DHS on enabling into the cellular system so that doing an emergency incident, you know, such as like 9/11 then certain users, you know, Federal and public safety users can get priority access on the commercial network. So that is a program——

Chairman BARTON. That could be done?

Mr. MULETA. And it has been done for some commercial networks. And it is in the process of being rolled out. You know, it takes a long time to, sort of, get all the procedures right. But we know of at least one national carrier that has put it in place, and others are in the process of considering it and trying to implement that.

Mr. GRUBE. Mr. Chairman, if I could add some comments to that, when we take a look at the interoperability, you know, it is all about process, planning and platform, platform being spectrum standards. When you look at the different levels of interoperability from basic just sharing radios to level 6 that we talk about, which is a common standard, the common digital P25 standard that law enforcement has endorsed adds encryption very easily.

I mean, that was one of the things that the users said they wanted when that standard was devised is the ability to easily encrypt it to a high level of encryption so that scanners, you know, from the department store are not going to receive sensitive information.

That is available. So as we talk about interoperability in, No. 1, joining together the people that need to talk to each other when they need to.

No. 2 is just giving them day-to-day better operation when they are just working within their jurisdiction. And that might mean transmitting a mode that others can't listen to the sensitive information. So as we do move forward with interoperability as the agencies do move toward the digital standard, P25, they will have that easy ability to add encryption so that their messages are not received by others.

Mr. BOYD. If I can add a couple of cautions, though, that we need to remember here, while P25 will allow encryption and that is built into the standard for P25, many of the PAP systems currently that are available that we are going to have to use for some time aren't very robust and have a difficult time in handling encryption when you begin to try to gather systems that aren't all P25 compatible. That is the first issue.

And Mr. Stupak addressed, I think, in part the issue that it is going to take some time for some of these other systems that are going to have to be included to make that transition. So thinking, planning—I think the point that Mr. Grube makes that is really important is prior planning. And that is that the organizations on the ground have to think out ahead of time both what they need to encrypt and where it is going to go.

The second piece of it that is important to recall is that as we think about things like reducing the size of channels, there is an overhead associated with encryption. And so, we will have to consider the robustness of the encryption algorithms as we make these decisions.

Chairman BARTON. Okay.

Thank you, Mr. Chairman.

Mr. UPTON. Thank you.

Ms. McCarthy?

Ms. MCCARTHY. Thank you, Mr. Chairman. I want to thank you, Mr. Chairman. I think this hearing and the panelists who have just testified has been one of the most realistic, down to earth presentations on the matter that we have had. And I appreciate your pursuing this because I know you know how important it is that we address this.

And we in the Congress have put out these great expectations through our earlier legislation, but we have never really come to grips with our role in carrying out the things that need to change to make it able for you, Chief LeGrande to accomplish what you so desire in the testimony you have before us today. So this whole broadband transfer issue, I would love it if you have some further wisdom to share on that.

I believe that, based on your testimony and what I have learned over the past month and years, is really critical to public safety that transfer has to happen. We have had hearings, but we have really taken no action as a Congress, as a committee to address it. But you lay out in your testimony all the reasons why it is necessary for our first responders to do their job adequately. And you have convinced me that it is now time for the Congress to act.

And the funds needed for digital equipment so that wherever they go they have got the high resolution video, whatever first responders need we haven't paid for, we haven't funded. So the Congress again has not done what is needed to make sure our first responders have the equipment that is interoperable.

I come from a community, Kansas City, where I have got two States kind of like the Virginia, Maryland, and you have got the District of Columbia to boot. But I understand the whole question for first responders because that dilemma exists in the greater Kansas City area, not just Kansas and Missouri and the river, but even within communities.

In my little community of Independence, Missouri, Harry Truman's hometown, the police and fire when they were trying to help with a dramatic ice storm we had a couple of winters ago couldn't communicate on their equipment to go in and help each other on-site. They ended up using cell phones. But in a terrorist attack, that is not a very good way to go.

So I guess my comment is I want to thank each and every one of you for reminding us today that we have a role to play in this in the Congress and we ought to be about it. But if you have some further thoughts to share, I would welcome them at this time in what little time I have left.

Mr. LeGrande. I thank you for your comments. We really appreciate that. I just want to clarify my title. I am not a first responder. I am am a part of the technology organization, although I do appreciate the compliment.

Broadband is clearly the next thing for, not only first responders, but it is the Nation. In meeting with, not only within the District of Columbia, our own MPD and fire, they have provided us with very stringent requirements on what they would like to be able to accomplish in order to meet the threat that exists. This system that we have already started to deploy on a pilot basis we already have reached out to our Federal partners to create interoperability with them, the U.S. Park Police and also the U.S. Capitol Police.

In fact, we can demonstrate within our current configuration a video feed from a U.S. Park Police helicopter from that helicopter that feed going into the FBI over to the MPD headquarters and out through our wireless network. We can show you where a first responder in the Capitol Hill region would be able to receive that feed via an Ipac computer, which is a small pocket computer or a laptop computer that, of course, is ruggedized.

This type of increase in capability for our first responders is exactly what they need. We have a threat that is multiplying exponentially, and we have to increase our ability to survey or to provide surveillance systems so that our first responders aren't completely tapped. They still have a domestic responsibility in addition to the new international threat that is been added to us. So these types of increases in technology will really help them to address that need. So we welcome the opportunity to present this much further to this committee, both in the legislation that is already included in our testimony, but also in demonstrations that we are capable of performing now.

Ms. McCarthy. Thank you. Would anyone else like to comment?

Mr. Grube. Yes, I would.

Ms. MCCARTHY. Thank you, Mr. Grube.

Mr. GRUBE. Thank you for your kind words. And it is a pleasure to be here. And I wanted to follow-up on the spectrum issue and tie that back to the applications that the public safety people seek. My team at Motorola has been doing trials for several years of what—to be compared to the last 65 years of two-way voice communications for public safety as breakthrough, revolutionary step change.

And back in 1997 when together, you know, we said let's target 100 megahertz, you know, for the public safety community and started making improvements toward that and most recently with the 4.9 gigahertz band, at that time, of course, pre-9/11, we didn't really understand the total picture yet.

And I think, you know, as leadership operations here in the district and other agencies that we have trialed broadband technology with, we have come to learn that not only will technology like this make a difference when these special events happen, but they can use it day-to-day in their operation to be safer, to be more efficient.

Some of the sound bytes that I have received firsthand from some of the public safety people trialing this technology is, "don't take it away. I feel safer when that broadband's streaming video technology is in my squad car to send an image of that traffic stop when I am out in the middle of nowhere sending that back to the dispatch center or to my partners so they can watch my back." I mean, these are the kind of words that they are telling us.

So we have learned, I guess, just recently in the last couple of years and since post-9/11 that, in addition to local broadband spectrum allocations of 4.9 gigahertz in addition to the high-speed data for Internet browsing, some simple video, limited capacity at 700 in the 24 megahertz there is a compelling need—and this is what the system in the district here is showing everyone for what I will call—and I will use my words very carefully—wide area broadband spectrum.

The 700 megahertz band is absolutely a sweet spot in terms of economics to bring that type of technology which is now coming into the industry, not only, you know, from companies like Motorola, but the entire industry.

A few weeks ago, the FCC had a very nice get-together on the wireless broadband topic. I served as a panelist. And there certainly was a common thread that the industry was saying. And this is not only for public safety. And I was there talking about that in addition to the consumer world.

But they are saying there is this rich spectrum here. There is new technology. We could deploy broadband, you know, for the industry. And we could deploy broadband for public safety. So the key message is the technology is here, the needs are now realized. And I think this additional spectrum that the district is talking about at 700 megahertz would be a wonderful thing to really go after and help enable as part of the platform for the public safety people.

Ms. MCCARTHY. Thank you very much.

Thank you, Mr. Chairman. I am glad I waived my opening remarks.

Mr. UPTON. You are lucky. I might just say before I yield to Mr. Bass that there are a number of us from this panel that actually witnessed that video transfer in Chicago. I think Mr. Engel was with us that day. I think Mr. Bass and Mr. Terry were there that day. And it is nice to hear as we listen to your testimony, Mr. LeGrande, that it is actually now that we are seeing it come into the field versus just a demonstration.

Mr. LeGRANDE. Thank you, Mr. Chairman.

Mr. UPTON. Mr. Bass?

Mr. BASS. Thank you, Mr. Chairman. I was rather surprised to hear that the system that you installed in the district cost $42 million. I am not saying it is high or low, but it is certainly a lot of money. And wondering about the question as to whether or not as we plan this interoperability and these new communications systems that we really have the best, most modern, diverse systems that we can possibly get.

I recall when I was on the Transportation Committee the FAA was authorized to spend an enormous amount of money on a new radar system which ended up being obsolete before it even arrived at the FAA centers, as I recall. And it cost many billions of dollars.

In the course of examining these systems, are you looking—let me start again. We had a hearing the other day in this committee in which we looked at the most unbelievably interesting new concepts for broadband communication and so forth.

And it has been my experience when you get into various sectors of government, the police or the fire departments or, you know, law enforcement, FAA, other agencies, they tend not to look outside of their own existing technologies, how they have always communicated, two-way radios with a microphone over your shoulder and so forth.

The fact is that a cell phone with a little television screen on it is probably not a bad way to communicate. Or it might be used as a basis upon which the agency—in this case, homeland security— looks at entirely new mechanisms and technologies outside of traditional radio communication, which provide by their very definition interoperability maybe a lot cheaper and with the ability to implement before digital transition occurs.

I am just curious if anybody in the panel has thought about this or, first of all, understands my question, but second, has thought about or has observations on whether the decisionmakers here are really looking at the big picture and trying to come up with a mission or a plan that doesn't get obsolete before—isn't obsolete before it is implemented.

Mr. LeGRANDE. Mr. Bass, first I would like to begin by agreeing with you that, yes, we should evaluate in parallel to addressing the urgent need for our first responders today, in parallel to that effort, you really evaluate emerging technologies and possibilities in the future. Currently I don't believe there exists a commercially available solution that we could quickly move to because the threat is so real and our needs to tie our first responders together is so important right now and time is of the essence.

As Mr. Stupak mentioned earlier, we didn't have the opportunity, nor would I suggest that other public safety organizations had the opportunity to move that quickly to the other solutions that exist.

A maturity needs to occur in moving to those solutions. And that is why we are piloting our broadband network within the district first.

We are going to run that pilot over a year as an experimental license that was already approved by the FCC for broadband communications. We are going to test security, reliability, maintainability. And those are some of the just varied components that we have to go into that maturity model before you can actually deploy these networks operationally.

The last thing we want to do is move quickly to either a commercially available solution or build our own solution based on commercial technologies without testing it out thoroughly because then we would run into a much larger problem. And so, what my suggestion to you, sir, is that first you have to start solving your short-term problem but in parallel work on the long-term solution that could take you to another place.

Mr. BOYD. I would like to add some——

Mr. BASS. Yes, before others respond, can I just add one other part to the question? Is there communication between homeland security and the military so that you guys know how the communication systems are working now with soldiers in Iraq, for example, and how well the interoperability issues that exist there and so on?

Mr. BOYD. In fact, that opens up an interesting question that I think you need to address in understanding the field. Now the first one is that I am retired from the United States Army. We are, in fact, working directly now with the Department of Defense with Assistant Secretary of Defense Paul McHale's office. We have worked with the National Guard, and the National Guard participates as a part of our executive council. And I have gone to address the Defense Science Board on a number of occasions.

There are a number of things you have to remember. And the defense model is an interesting one to remember. The Defense Department first became interested in becoming interoperable some 10 years before I was commissioned a second lieutenant in the United States Army. It did work diligently at that. And today, some 12 years after I retired from the Army, they are almost interoperable. That is four services in a single department that is funded essentially by a single committee.

In this community, we are looking at 50,000 independent agencies who are funded by the city councils, county commissions, by the states and others. These are organizations which, for the most part, have communication systems which have a life cycle that is on the order of 30 or 40 years. Within the technology life cycle, it is 18 to 24 months.

So part of the difficulty we have is how do you make sure that you don't leave behind the community that can't afford to upgrade its system but needs interoperability now, too. And so, how do you bring these together?

It is, I think, tempting to imagine that there is a single standard and a single technology that is going to solve the problem. There is not. The reality is that we will always have multiple systems over a period of time, even when we arrive at fairly common standards because we don't want to stop the innovation of the technology.

And so, you may have newer technologies developing that will always require us to think ahead about how we are going to tie them together. Which is why the approach—we took a batch of systems—you need to think about routine communications and emergency communications. The things that you use routinely need to be the same things you are going to use in an emergency, otherwise they not only won't know how to use them—the military operates on the same basis—they may not even know where they are stored.

So you need to use the same kinds of systems in both cases. And you need to understand which parts of your routine communications you can off-load onto, for example, the commercial structures. In fact, one of the things we are working with local law enforcement with is to help them to understand where they can build in as part of their plan some of the commercial infrastructure.

But it is important to understand that the cellular and the public switch telephone network, that is the wired network, are built only for a capacity that is about 10 percent over the normal capacity, which is why during rush hour you frequently can't get a call on a cell phone. And in an emergency, almost by definition, those systems are overwhelmed almost immediately.

So there has to be an emergency foundation that the public safety community can fall back on. And we have to understand that while we don't want to take 45 years as Defense did to get to interoperability, we also are not going to get there in 1 or 2 years.

Mr. BASS. I yield back, Mr. Chairman.

Mr. MULETA. Mr. Engel?

Mr. UPTON. Do you want to respond to that question?

Mr. MULETA. Yes, I think it is an important issue, and I just wanted to give you a couple of points that I think are very important. One is the kind of requirements development that DHS through SAFECOM is doing right now will really help. A lot of times the tradition has been in the public safety environment that you, sort of, take the equipment as is, that, sort of, technology is leading as opposed to the requirements leading. So I think that is a very important step in terms of the cultural change that is going on.

In terms of what the FCC has been doing, in various proceedings that we have had regarding public safety radios, we have tried to incorporate standards-based approach into the use of the spectrum. So, for example, the 700 megahertz, the process that we used there included regional planning, and it also included the adoption of project 25 into what public safety can use there. In terms of the intelligent transport system, which is something we work cooperatively on with FAA which is about vehicular accident effectively systems that are just put on a terrestrial basis that feed information or allow communications for DOT and DOT-sponsored organizations. We again used various standards-based technology to be adopted in that order.

So as, you know, car makers and as people are developing radio systems to use that, they are, you know, effectively using off-the-shelf components. And they are riding the curve, the technology curve and the cost curve. But the commercial world experiences a lot faster than the public safety has traditionally.

In 4.9 gigahertz band, which is 50 megahertz dedicated to public safety, again, we used basically equipment, you know, sort of, standards that are used in the 5.8 gigahertz, which is where a new allocation of Y5 is being placed. And so, other things that we are doing are cognitive radios, effectively helping the development of technology that allows radios to, sort of, flexibly move from one band to another.

And so, in the case of an emergency if everything is tapped out in spectrum a, you can move to spectrum band b. So the FCC has a proceeding on this. So a lot of what we are working on is trying to get to standards-based solutions and embed them into the regularity model that we are using so that people won't be caught—you know, actual operators like Mr. LeGrande won't be caught short as technology moves or the cost curve declines significantly in the commercial world.

Mr. GRUBE. Mr. Chairman, I have another comment.

Mr. UPTON. Just very quickly. I stopped the clock.

Mr. GRUBE. Congressman Bass, Terry and Chairman Upton, thanks for looking at the high-speed data pilot that we had in Chicago. And I think that, you know, one of the take-aways when public safety looks at deploying their own private networks is your original question does come up a lot. And that is can we use a consumer-based carrier network for the first responders. And they do use them from time to time.

But if you look at the economics of a carrier system, it is driven by putting just the coverage that you need where the highest population of people is. And it may not be the third sub-basement where the firefighter has to go or the police officer has to go. It may not be the far reaches of the county where the State patrol officer has to go. So coverage is always a key thing. And that is one of the reasons the private networks are here.

And I think that—and one of your other questions dealt with the technology that is here and questioning is this here to stay. I think the basic two-way voice—if you talk to the people who carry guns and hoses, their primary need is to push the button and to be instantly heard by someone at the end of that radio communications path. Others around them that are supporting the scene or supporting, directing what they do, those are the ones that also need that capability plus all of the richness that we have talked about in terms of video and Internet access and those things.

And again, if we take a look at the carrier networks, we learned a big lesson during the major blackout last year in the Northeast. Half of—about half of the carrier cellular sites were down. And that is because economics just don't motivate the carriers, you know, to design around that. Whereas if you look at the State of Michigan, their system was fully operational. Everything was taken into account for, not only capacity spikes, but also in terms of power outages.

Mr. UPTON. Thank you.

Mr. Engel?

Mr. ENGEL. Well, thank you very much, Mr. Chairman, for calling this hearing. This proves time and time again why this is a great subcommittee and why we are always on the cutting edge of things that are really on people's minds.

I know Mr. Stupak spoke about September 11 and the problem in my home State of New York about the policemen and firemen talking to each other. But, you know, just Monday, a couple of days ago, I was meeting with police, fire and EMS officials in Ramapo in Rockland County which is a northern suburb of the city of New York in my district. And we were talking about rail security.

And the fire inspector there said to me, "You know, our biggest problem is if we have an emergency, the first cop and first firefighter on the scene can't talk to each other on their radios." So this is obviously something that we are still hearing from all across the country.

I wanted to mention before I asked my questions that I have worked on a bill, a bipartisan bill with my colleagues on this committee, Mr. Stupak and Mr. Fossella, to provide the funding that our local first responders need. It is called the Public Safety and Interoperability Implementation Act, which is H.R. 3370. And what it does is it would reserve a portion of future spectrum auction revenues and place them in a trust fund for helping State and localities in paying for these new systems.

So, Dr. Boyd, when you mentioned that smaller communities, smaller areas and towns really don't have the money, we would envision that if this bill were to be passed and implemented that that would be a way of providing those kinds of funds.

Mr. Chairman, I want to ask unanimous consent. I have with me a letter from the county executive of Rockland County, Scott Vanderhoef, requesting assistance in obtaining Federal funds for the complete overhaul and upgrade of emergency communications in Rockland. Even though he was my opponent in 2002, I will certainly be helping him because he is right. And I would just ask unanimous consent to enter into the record his letter as an example of what our localities are facing when trying to afford an interoperable system. I will ask unanimous consent for that.

Mr. UPTON. Without objection.

[The letter follows:]

COUNTY OF ROCKLAND
OFFICE OF THE COUNTY EXECUTIVE
NEW CITY, NEW YORK 10956
June 3, 2004

The Honorable ELIOT L. ENGEL
The United States House of Representatives
2264 Rayburn House Office Building
Washington, DC 20515-3217

DEAR CONGRESSMAN ENGEL: This letter is to request your assistance with a matter of great importance to the citizens of Rockland County.

Since the tragic events of September 11, 2001, more than $23 billion have been appropriated to help States, municipalities, and first responders improve preparedness for future acts of terrorism or other emergencies. And since March 2003 the Department of Homeland Security has specifically helped first responders prevent, prepare for, and respond to acts of terrorism.

As you know, Rockland County is located a mere 15 miles north of the George Washington Bridge. Along with its close proximity to New York City, Rockland lies within the response area for the Indian Point Nuclear Power Center. Interstate highways, rail traffic and Hudson River access pose certain challenges in protecting our most vulnerable and valuable assets.

I thank you for introducing the Public Safety Interoperability Implementation Act in order to focus the resources of the federal government on those areas that are most vulnerable. Homeland Security Interoperable funds could help facilitate the implementation of Phase II of Rockland County's Public Safety System Communications Project. This project is designed to allow our fire, police and ambulance responders to effectively communicate with each other even under the most challenging of circumstances.

I am sure you can appreciate that this project should be considered one of the highest of priorities among all government levels. Terrorists are not arbitrary in their selection of targets and some of the region's most vulnerable sites and communities lie within the 17th Congressional District.

Therefore, I present the attached project outline in hopes of accessing federal funds for this invaluable project.

Thank you for your consideration and continued advocacy on behalf of the citizens of Rockland County.

Very truly yours,

C. SCOTT VANDERHOEF
County Executive

Rockland County
Public-Safety Communications System
Phased Roll-Out

Estimated Total Cost: $27.2 to $28.2 Million

Phase I	Phase II	Phase III	Phase IV	Phase V	Total Project
12th Month	24th Month	24th Month	36th Month	48th Month	48th Month
$3.7M to $4.2M	$10.2M (a)	$3.5M (b)	$5.0M (c)	$4.8M (d)	$27.2M to $28.2M
Contract Awards Radio Infrastructure Site Development Microwave	Complete Site Development Complete Microwave Install Infrastructure (5-Channel) System Testing	Fire Switch-over	EMS Switch-over System Upgrade to 13 Channels	Police Switch-over	Fire, EMS & Police 13-Channel Digital Trunked Radio System

(a)   Includes Contingency $1M
(b)   Includes Contingency $0.6M
(c)   Includes Hospital and Ambulance Companies Control Station
(d)   Includes Remote Consoles, but not interconnection

Mr. ENGEL. Thank you.

Mr. LeGrande, I appreciate you have put so much time working into the engineering design of the DC system. Obviously we are all concerned. We work here. And DC faces many unique problems caused by overlapping agencies and jurisdictions.

And I applaud you for making interoperability with the Metro subway system a priority. I am wondering, though, how much training is going on so that Metro workers, police, fire and EMS, know what is going on. How many hours of training does a Metro worker receive? Do they receive training for this sort of thing?

Mr. LeGRANDE. Well, the system that we put in for our fire department doesn't require any additional training to use it in the Metro system. It is seamless, and in just the same way they use it above ground, they use it below ground and while riding on the trains. So no additional training is needed there.

Mr. ENGEL. What about the Metro worker?

Mr. LeGRANDE. When you speak of the Metro workers, they have their own communications systems and their own set of training that is gone on. Their systems have been in place for quite some time, so I don't think there was any additional training needed for them. ENGEL: All right. Now you mentioned that your system has interoperability capabilities with the Capitol Police. Am I correct? You said that they did.

Mr. LeGrande. For the U.S. Capitol Police, who I said is we put the capabilities in, yes, to have interoperability with them. And we need to work out those standard operating procedures with them to facilitate that communication.

Mr. Engel. Okay. Obviously Capitol South Metro is adjacent to the Cannon Building, and our police, the Capitol Police, would respond to an incident there. Has there ever been a drill held at that location?

Mr. LeGrande. None that I am aware of. I am here representing the technology that we put in. The actual operations of the police—that would really have to come from the MPD. I can find out the answer to that question for you, though.

Mr. Engel. Okay. And would you know how often drills are held?

Mr. LeGrande. No.

Mr. Engel. Do you have any idea if that includes the Metro Police, fire and EMS?

Mr. LeGrande. I wouldn't know the answer to those questions, no, sir.

Mr. Engel. Okay. If you could find that out for me, I would appreciate it.

Mr. LeGrande. Okay.

Mr. Engel. Okay. Thank you.

Mr. Grube, I want to thank you for the information. It was very extensive. There have been some efforts to immediately move to TV stations operating on channel 63, 64, 68 and 69 off. And I am wondering if you could give me some information about that.

Mr. Grube. Well, one of the—there are several methods. And one is to simply relocate the channels, the TV stations that are in those channels down to a lower channel but still transmit in the analogue mode. That is one way. And some stations have applied for waivers to do that, I understand.

Another way is to have them move in the move to the digital mode when they move out of that band and to provide the 3 percent of the households, according to the independent analysis that we did, with digital to analogue converter boxes so that they could still—that those consumers could still continue to use the analogue TV equipment that they have today. Those are a couple of the methods.

Mr. Engel. Well, we have a problem in New York. I don't know if you are aware that there are adjacent TV channels that are in use. Would those TV signals cause interference? And would we have to shut down those adjacent channels as well?

Mr. Grube. They can't—yes, they should be included in the analysis because, you know, a 5 megawatt transmitter spectrally next to, you know, a poor, little homeland security radio could be an issue. So I think that has been included in the analysis that we have done. And it is very important to consider those stations as well.

Mr. ENGEL. Because, for instance, in the New York City metropolitan area, channel 67 is Univision on Long Island. And channel 68 is Univision in Newark, New Jersey. And both serve the New York City area. And during an emergency, obviously Spanish-speaking people turn to Spanish language news. Thus there is a competing safety concern as well. So I am happy that it is included as well.

Thank you, Mr. Chairman.

Mr. UPTON. Mr. Fossella?

Mr. FOSSELLA. Thank you, Mr. Chairman, gentlemen.

For Mr. Muleta, thank you for your testimony. Currently FCC is looking at giving public safety additional spectrum and other band widths. Is that correct?

Mr. MULETA. Yes.

Mr. FOSSELLA. Some have suggested that additional spectrum at the 700 megahertz band would be more useful for offering interoperable broadband services. Do you agree with this?

Mr. MULETA. Seven hundred megahertz has, as we have talked about, good propagation characteristics. And there is already 24 megahertz that is been allocated at 700 megahertz for use by public safety.

Mr. FOSSELLA. So in light of the ongoing discussion, do you think having a larger block of spectrum in that single band width to allow for more efficient use of spectrum makes the interoperability easier to achieve than what is currently proposed? Or as stated otherwise, is a single block of spectrum better than the fragmentation of spectrum and other bandwidths?

Mr. MULETA. Well, I think there is already an allocation of the 700 megahertz. There is 24 megahertz that is been allocated. There has been a national plan put in place, a unified standard for the technology in project 25.

I think the issues that, you know—the solution to the problem that we have today on interoperability have more to do with planning, coordination, communication and, you know, sort of, actual deploying them, getting the dollars out to get the systems up and also obviously the fact that that spectrum is encumbered with the broadcasters.

So additional spectrum—you know, additional blocks of spectrum, I think, would always be useful in any communications context, whether public safety or any other application. I don't think the current—that, you know, today an additional block will get us, sort of, the uniform interoperability that we are all looking for because I think those require communication coordination and planning more so than additional spectrum, from my perspective.

Mr. FOSSELLA. Okay. Thank you.

Mr. LEGRANDE. Could I add a response to that question, sir? From our perspective, the 24, while it has been allocated to public safety, the current configuration of the 24 does not allow for wide area broadband use. So on the issue of interoperability, I don't disagree. But on the issue of deploying wireless broadband technologies in the current 24 megahertz configuration, we cannot do that over a wide area.

Mr. FOSSELLA. Mr. Muleta, in light of that, do you have any comment?

Mr. MULETA. Yes, I think there have been several different requests that have come through for additional blocks of spectrum. I think by statute Congress has—you know, 24 megahertz was allotted to public safety. The additional spectrum was for commercial uses for auction at various times. And I think there are, you know, countervailing tradeoffs.

I think again the key is to get interoperability into the hands of the folks today as quickly as possible. As Mr. LeGrande so aptly put it, there needs to be interoperability then interoperability and laying an additional amount of spectrum, although it is useful as part of our planning process, we need to—I think it needs to go through the whole process of what the statute has asked FCC to do, which is to design it for—this additional spectrum for commercial uses.

Mr. LeGRANDE. I would just disagree with the characterization that it will be helpful. It is needed for public safety to have wireless broadband technologies available in order to address the threat that exists to our country. So I totally agree that the 24 megahertz in the current interoperability plans and the efforts that we are all making here. But it is important that we also move right now to providing that ability for our first responders.

Mr. GRUBE. Could I add a comment? You know, I think the original Biswick Report did not contemplate at that time wide area broadband wireless. And what we have been discovering in the last few years is that there is a real need for this, one that we didn't see before so that when we talk about the 700 megahertz band, presumably we are going to fix this, we are going to clear it. That is a voice interoperability, the basic data, not the wide area broadband. And its proximity to the 800 megahertz band in total makes a very nice economical way for the industry, like Motorola, to provide product across those 2 adjacent bands.

But in addition to that, to solve the needs that we have been discovering together about broadband, we feel strongly that an additional allocation is required. And since public safety is already bracketed by the 24 megahertz below and the 800 megahertz above, it makes a lot of sense to consider public safety and Federal broadband, wide area broadband interoperability in this 30 megahertz that we are talking about.

And I think that the economics will help determine this as well because from a propagation point of view, when you take a look at the cost to go build a 700 megahertz wide area broadband system for a relatively few number of users—and I say that about public safety because if you look at the user density per square mile for public safety, it is a different equation, probably by a factor of 100 relative to the consumer world.

And so, if you are a carrier and you are contemplating the bands, you are probably already talking about smaller cells that spectrum such as the 2.1 gigahertz band or others up in that area would be very attractive in terms of spectrum that they would like and pay money through an auction. So I think that we have to together look at those options for the carriers that want to create a business there and really take as a priority the sweet spot, if you will, at 700 megahertz for public safety.

Mr. BOYD. Just as a quick note, the public safety community will tell you that right now they have a lot of priorities, but two fundamental priorities. One of them is the elimination of the interference problem on 800 megahertz. And the other is additional spectrum. And they will tell you their priority concern, while they are interested in all kinds of communications, is always going to be voice.

Mr. UPTON. Mr. Stupak has a couple of additional questions.

Mr. STUPAK. Thank you, Mr. Chairman.

If the priority and if their basic system is always voice, then what is standing in the way of just going and getting voice out there so they can talk to each other? I mean, they all have voice now as a basic component. They are on different frequencies. And we have the technology that already exists and has been around for a long time to allow different frequencies to speak to each other. So why don't we as a first step, almost 3 years after 9/11, just do that part so they can just talk to each other? Can that be done?

Mr. BOYD. A number of efforts were already being done to do that. In fact, most of the major urban areas, even beyond the U.S. areas that are defined by ODP have, in fact, patch technologies and patch devices in place. But even these require some time to put in place because it is not just a technical problem. Part of it is a technical issue and, of course, the costs associated with training and planning that goes with it. But the other part of it is very much a cultural issue.

In 1993 when I first put together an interoperability initiative in this case, we wanted to try to allow all of the agencies, Federal, State and local in San Diego County to communicate with each other. And we used a fairly primitive switching technology, nothing as good as exists right now.

Mr. STUPAK. Sure. Right.

Mr. BOYD. It took us 30 days to implement the technology with us paying for it. It took 2 years to get all the agencies to agree to play a role.

Mr. STUPAK. But don't you think that is all changed since 9/11? I mean, not all changed, but has really lessened since 9/11 and the different needs and different things that we are asking them to do from the Federal Government from a terrorism point of view?

Mr. BOYD. I think it is much easier to get people on the same sheet of music.

Mr. STUPAK. Sure.

Mr. BOYD. But you still have to take the time to do that.

Mr. STUPAK. Sure.

Mr. BOYD. And that means that you need to set up a process that brings in—the way we talk about it is that you need to set up a process that lets the local guy have a serious incentive, that makes him want to be part of this. There is a tendency to try to force things from the top down, whether it is from the Federal level or the State level.

Mr. STUPAK. Right, I agree.

Mr. BOYD. And you cannot do that in this community.

Mr. STUPAK. But the incentive for doing it is just basic safety. It is basic safety. It is more and more municipalities are going to one-person cars which they did not do before. And I think we can see it from a number of examples. Going back to San Diego there, if you could do the technology in 30 days, what was the cost then?

Mr. BOYD. Well, at that time, we used an existing Navy switch panel to do that.

Mr. STUPAK. Right.

Mr. BOYD. And it essentially was a manual patch, and operators sat there and tied them together.

Mr. STUPAK. Right.

Mr. BOYD. It is not like they put them now. I would say we probably invested to do that in that county just to do the technology about a half a million dollars. That did not support—understand it was—it took more to do that.

Mr. STUPAK. Right. Yes.

Mr. BOYD. But that doesn't support the continuing training and the manning of the system and so on.

Mr. STUPAK. Right. But now with a lot of jurisdictions going to 911 and emergency 911, E-911, it is a lot easier to do this now, to get the coordination and jurisdictions down under at least one call center.

Mr. MULETA. If I could answer that question——

Mr. STUPAK. Yes, go ahead.

Mr. MULETA. Well, I think most jurisdictions, public safety answering points which deal with E-911 are different than the public safety radio systems which are usually allocated. You know, the public safety answering point is——

Mr. STUPAK. Sure, but doesn't the 911 also not only answer but also dispatches?

Mr. MULETA. It dispatches, but it goes through, I believe—let's say it dispatches directly to the metropolitan police department which then makes a decision to go to. So there isn't a direct link to the officer on the ground. It has actually to go through——

Mr. STUPAK. Well, we in the rural areas still do it correctly. We don't have a metropolitan to go through. So we are usually receiving and directing right back out to the cars. And that is why interoperability is so important when you are dispatching to a State, local or county sheriff. I mean, it is all got to be the same.

Mr. Boyd, if I can go back, of the $4.4 or so in Department of Homeland Security grants that have gone out the last 2 years, do you know how much has been dedicated to interoperability?

Mr. BOYD. That is a really tough question. And I can explain why. We know in the case of the interoperability grants in COPS and FEMA last year, about $75 million in each agency. We know that that was interoperability money.

Mr. STUPAK. Right.

Mr. BOYD. And, in fact, we participated in helping to set up the selection process to do that. And we know that the $85 million in the COPS Office this year is interoperability. Most of the rest of the money, however, is block grant money which goes to the states.

Mr. STUPAK. Correct. Correct.

Mr. BOYD. And as you know, once it gets to the State, the State then can provide it to localities for any of a series of authorized uses. The states aren't obligated to report back on how much of that is actually used for interoperability, for example.

Mr. STUPAK. Okay. Yes, last week I was going to do an amendment on the House floor, and Chairman Rogers thought that you might be able to come up with those figures because I didn't want to waste the money. I would rather see the money go into the interoperability——

Mr. BOYD. In fact, we may be able to help you. I am not sure about——

Mr. STUPAK. But even in Michigan—we called Michigan. They have received about $120 million, and they could not tell me how much was interoperability. Hopefully there will be some way we can focus on this in the next few months because we are looking at a huge price tag. And we talk a lot about Federal Government having to take leadership. And we certainly do have to put the money forward for this. So we would be interested to see what has gone in there and how much it is going to take.

Mr. BOYD. Yes, one of the pieces that may help you with that, that we hope will help you with that because we think it is crucial to what we are doing——

Mr. STUPAK. Sure.

Mr. BOYD. [continuing] is a baseline survey which we are initiating right now. Now this survey will take probably about a year to complete. But we want to try to get a picture of what the level of interoperability is. And you can't go to any place now. You can't go to a data base.

Mr. STUPAK. Right.

Mr. BOYD. You can't go to a source and say, "What is the level." So we want to do a really well-designed survey to get a picture of what that baseline is because we need to bounce that against the statement of requirements we have just produced, figure out what the gap is and then we can give people realistic estimates of what it is going to take to move to interoperability.

Mr. STUPAK. Yes, and you might want to take a look at Michigan. They have been one of the leaders in it. But still, even with your new system—they just did it statewide to the State police. They still have 1,000 public safety agencies in the State still not tied into it and still don't have the interoperability. So that is a good place to start because they just completed theirs last year.

Mr. BOYD. Yes, sir. I spent some time working with Mike Robinson. I am familiar with the system.

Mr. STUPAK. Good.

Mr. UPTON. Thank you. I would just comment on the fact that was raised about E-911. I had the occasion to make an E-911 call the other day at 6:30 in the morning. And the system here works. I was on the bridge coming from Virginia to DC, saw a terrible accident and called in. I was immediately patched to PSAP somewhere in DC and identified that it was Arlington County. Or the accident actually was. And they responded. So the system is working.

And this is a very important subject, that one as well as interoperability. I am pleased to see that things seem to be in place and moving forward in the right direction. But obviously the rest of the country, whether it be in Michigan or other places are on the ball as well.

And just as I talk about E-911, I lament the fact that this sub-committee and committee in Congress, thanks to bipartisan help in a major way passed a very good E-911 bill. And it is still languishing more than a year later in the other body, as they like to say. Some of us like to say the lower body, but we won't say that.

But again, I appreciate all of your work. This is a very important topic, not only for us, but for the country. And we appreciate your leadership. We look forward to working with you as we move the ball down the field.

This hearing is now adjourned.

[Whereupon, at 3:20 p.m., the subcommittee was adjourned.]

○